Daniel Giraud Elliot

The Wild Fowl of the United States and British Possessions

Or, The Swan, Geese, Ducks, and Mergansers of North America

Daniel Giraud Elliot

The Wild Fowl of the United States and British Possessions

Or, The Swan, Geese, Ducks, and Mergansers of North America

ISBN/EAN: 9783337192389

Printed in Europe, USA, Canada, Australia, Japan

Cover: Foto ©ninafisch / pixelio.de

More available books at **www.hansebooks.com**

THE
WILD FOWL

OF THE

UNITED STATES

AND

BRITISH POSSESSIONS

OR THE

SWAN, GEESE, DUCKS, AND MERGANSERS

OF

NORTH AMERICA

WITH ACCOUNTS OF THEIR HABITS, NESTING, MIGRA-
TIONS, AND DISPERSIONS, TOGETHER WITH DESCRIP-
TIONS OF THE ADULTS AND YOUNG, AND KEYS
FOR THE READY IDENTIFICATION OF THE SPECIES

*A book for the Sportsman, and for those desirous of knowing how to
distinguish these web-footed birds and to learn
their ways in their native wilds*

BY

DANIEL GIRAUD ELLIOT, F. R. S. E., ETC.

Ex-President of the American Ornithologists' Union

*Author of the New and Heretofore Unfigured Birds of North America; of
the Illustrated Monographs of the Ant Thrushes (editions 1863 and
1895), Grouse, Pheasants, Birds of Paradise, Hornbills, Cats,
etc.; of the Classification and Synopsis of the Tro-
chilidæ; of the Shore Birds, and Gallinaceous
Game Birds of North America; of
Wolf's Wild Animals, etc., etc.*

WITH SIXTY-THREE PLATES

NEW YORK

FRANCIS P. HARPER

1898

COPYRIGHT, 1898,
BY
FRANCIS P. HARPER.

THE AUTUMN FLIGHT.

From the strongholds of the North
When the Ice-King marches forth,
The Southern lands to harry with his host;
The fowl with clang and cry
Come speeding through the sky,
And steering for the shelters on our coast.

I hear the swish and swing
Of the fleetly moving wing,
I see the forms drawn faintly 'gainst the sky,
As the rush of feathered legions
From out the frozen regions,
Sail onward 'neath the silent stars on high.

Like a cloud that's borne along
By a mighty wind, and strong,
Then parting, disappears in vapor light,
They glide o'er lake and sea
O'er mountain, moor, and lea,
And, passing swiftly, vanish in the night.

They seek a sunny clime,
A land of blooms and thyme,
The tranquil surface round the southern Key;
A home of peace and rest
On the friendly water's breast,
Of lake, or flowing river, or the murmuring sea,
The gently heaving bosom of the sea.

PREFACE.

THE Swan, Geese, and Ducks naturally become the subjects for the third volume of what may be called the series, or trilogy of " North American Game Birds." While engaged upon this book, I felt that I was writing the history of a rapidly vanishing race, whose serried hosts, at a time not far distant, were spread over the entire length and breadth of the continent as they winged their swift flight in the annual migrations. But incessant persecution and unrestrained slaughter have been waged against these fowl, in all manner of ways: by killing the mated birds in the spring on their way to the breeding grounds, by robbing the nests, by murdering the young perhaps even unable to fly, and by continued shooting during their southern journey and in their winter residence—until to-day but a remnant is left of the myriad fowl that at one time fairly darkened the air with their mighty legions.

And although it is apparent to all, save those who will not see, that only a brief period can elapse, if the same conditions continue, before, like the buffalo, our Water Fowl will mostly disappear, yet little is done to save them from destruction, and the ruthless slaughter goes gayly on. Improved firearms of all kinds and devices of every sort to reduce their numbers, each one more dangerous and effective than its predecessor, are continually being introduced, and there is hardly a spot all over our broad land where a wild Duck or Goose can rest a moment in peace. From the time the birds leave the frozen Northland, until the survivors return to it again in the ensuing year, the hunted fowl run the gantlet of a nation in arms; and no sooner do they pass the boundaries of the land they seek in the spring for the purpose of reproduction, than the natives continue the slaughter of the birds until they depart for southern climes. Is it any wonder that their numbers are diminishing; is it not rather a wonder that so many are left? Doubtless these fowl are one of the important means for sustaining the lives of those who exist in Arctic solitudes, but the natives, before they were taught the white man's ways, carried on no war of extermination, and the number of the invading army did not diminish, as is proved by the myriads that once entered the United States every autumn. But now, provided with modern firearms, in place of the spear and the bow and arrow, the savages slay the birds not alone for their own consumption, but also to supply the demands of commerce and of fashion, while the eggs are collected by boatloads in order that certain pursuits may be made more profitable. By such mischievous methods the misguided inhabitants of the Arctic regions are destroying one of their own means of existence, and joining with civilized man in southern climes, to hasten the extermination of the race.

Few families of birds have more admirers than that of the Anatidæ, and in the early autumn the *Honk* of the Goose, or the whistling wings of the advancing army of Ducks, heard overhead at night as they arrive from the North, cause many an eye to glisten, and many a pulse to beat faster throughout the land. Duck-shooting has a host of votaries,—perhaps no kind of field sport has so many,—who follow it enthusiastically in spite of its often attendant hardships and exposures.

This volume is arranged on the same plan as those of the "Shore Birds" and "Gallinaceous Game Birds," now pretty familiar to my readers. The species, however, which are contained in this book are fairly well known to most people, at all events the males are; but as the females of different kinds often resemble each other closely, I have endeavored in the Keys, when necessary, to draw comparisons between them and call attention to their most marked characteristics. In the arrangement of the Family, occasionally in the selection of genera necessary to designate the different groups, and in certain cases also, in the choice of specific names, as well as in the general disposition of the species, I have found myself obliged to depart widely from the method adopted in the A. O. U. Check List, which seems in a great degree to have been constructed without sufficient consideration of the affinities the North American Anatidæ might possess to the exotic members of the Family. Of course no natural arrangement is possible, for none exists, but I have endeavored to bring together those groups which were most in accord and produce a proper order of succession, although fully aware that gaps occur.

No birds vary more, even if as much, in their relative dimensions, as do the members of this family. Not only is there great divergence among the species of a genus, but also even among those which are members of the same species. In fact it is not easy to find any two Ducks or Geese which are exactly alike in all their measurements. To ascertain how great these differences are, it is only necessary to consult Mr. Ridgway's "Manual," when it will be seen that for a large proportion of these birds an *average* measurement is given, instead of an EXACT one, and I have found so much variation existing that in many cases, when the dimension of a species is recorded,

I have been compelled to qualify it with the word, "about."

Bearing this fact in mind, therefore, I consider it most unwise and injudicious to create even a subspecies whose only character is that of size, especially when it is attempted to separate birds of different lands which are so exactly alike as not to be distinguished apart until the tape-line is applied, and even then the test fails at times, as they are often found to be of the same dimensions. It will be observed, then, that in certain cases I have not recognized such so-called subspecies or allied forms, believing that, should I do so, I would only confuse my reader and perplex any student conscientiously desirous of studying specific relationships. The fact that a species is found in Europe and America is no reason whatever that the specimens from the two hemispheres must be specifically, subspecifically, or in any other degree separable, simply because they come from different localities. Yet it would seem that in certain cases some writers were convinced that such must be the fact. A comparatively slight difference in size alone, however, is utterly unreliable as a distinguishing character, and should receive little consideration, save when accompanied by other and more important distinctions.

In the Appendix will be found Keys to the Subfamilies, Genera, and Species, and such critical remarks as more properly find there a place.

The Author has studied the Anatidæ for many years, and he has with but few exceptions met all the species mentioned in this book in their native wilds, and the accounts given of their habits are derived from his own observation. The majority nest in places not difficult of access, but for a history of the ways, in the breeding season, of the few species that are then found only in the far

Arctic regions, the Author has relied upon the naturalists who have had opportunities of observing them in those distant localities. The position and names of the feathers of the Wild Fowl do not differ from those of other birds, and they can readily be ascertained from the plate given in the " Shore Birds " which serves the purpose of an explanatory map. The drawings of a considerable number of the species were made by the Author at a time when he was contemplating another work on the Water Fowl, and these have been reduced by Mr. Edwin Sheppard to the proper dimensions for this book. Four are reduced copies of paintings made by the great artist Joseph Wolf, for the Author's work on the " Birds of North America." The remainder of the plates have been drawn by Mr. Sheppard, who illustrated the two previous books of this series, and these exhibit the same care and fidelity in their execution as characterize the plates in those volumes.

For the loan of specimens from which the drawings by Mr. Sheppard have been made, I am indebted to my friends Mr. R. Ridgway, Curator of Birds in the National Museum, Washington, and Mr. Witmer Stone, Curator of Ornithology in the Academy of Natural Sciences of Philadelphia, to whom I desire to express my thanks for their assistance.

In classifying the various groups of the Anatidæ it is of slight moment whether one begins with the so-called highest or lowest form, naturalists having not yet agreed upon this point, although it would seem advisable in the construction of a pyramid to begin at the bottom and not at the top. In the arrangement of the genera and species in this book, however, I have reversed the order in the Check List of the American Ornithologists' Union, because I desired to begin with the most important species of the Water Fowl, and therefore commence with the Swan instead of the Mergansers.

In this and the two preceding volumes have been included all the birds inhabiting North America, north of Mexico, which can be considered "Game," save perhaps the Rails, which by many are deemed worthy of being so classed. It is a noble list; one few countries of the globe can equal in importance and variety. For numerous reasons, not the least of which are the economic, these birds are a most valuable possession to the people of this land, to be protected with watchful care. Have we been faithful to our trust?

In the willful destruction of all our feathered creatures that has been permitted without restraint for a long period throughout North America, and which receives but little check to-day in some districts; in the lack of all intelligent treatment of them within our limits; and in the non-enforcement of laws passed for their protection, our birds (not only, alas! those entitled to the epithet of "game") are being carried rapidly onward toward extinction. Our wild quadrupeds, also, are fast disappearing. One, the grandest of all, is even now practically extinct, and unless stringent measures are soon taken and the laws for both their protection and for that of all other wild creatures rigidly enforced, waters without their beautiful, joyous tenants, and plains and forests despoiled of their graceful inhabitants, will bear silent but eloquent witness to the folly of a people unable to appreciate the valuable gifts Nature had bestowed upon them.

D. G. E.

TABLE OF CONTENTS.

	PAGE
THE AUTUMN FLIGHT,	v
PREFACE,	vii
LIST OF ILLUSTRATIONS,	xv
INTRODUCTION,	xvii
WHISTLING SWAN,	19
TRUMPETER SWAN,	28
WHOOPING SWAN,	31
BLUE GOOSE,	33
LESSER SNOW GOOSE,	35
GREATER SNOW GOOSE,	39
ROSS'S SNOW GOOSE,	43
WHITE-FRONTED GOOSE,	45
BEAN GOOSE,	50
EMPEROR GOOSE,	52
CANADA GOOSE,	57
HUTCHINS' GOOSE,	69
WHITE-CHEEKED GOOSE,	72
CACKLING GOOSE,	74
BARNACLE GOOSE,	78
BRANT GOOSE,	80
BLACK BRANT,	84
WOOD DUCK,	87
BLACK-BELLIED TREE DUCK,	92
FULVOUS TREE DUCK,	95
RUDDY SHELDRAKE,	97
MALLARD,	100
DUSKY DUCK,	106
FLORIDA DUSKY DUCK,	109
MOTTLED DUCK,	111
GADWALL,	113
EUROPEAN WIDGEON,	116
WIDGEON,	118
SPRIGTAIL,	122
BLUE-WINGED TEAL,	128

TABLE OF CONTENTS.

	PAGE
CINNAMON TEAL,	132
EUROPEAN TEAL,	134
GREEN-WINGED TEAL,	136
SHOVELER,	140
RUFOUS-CRESTED DUCK,	144
CANVAS BACK,	147
RED HEAD,	154
SCAUP DUCK,	160
LESSER SCAUP DUCK,	164
RINGED NECK DUCK,	169
LABRADOR DUCK,	172
GOLDEN EYE,	176
BARROW'S GOLDEN EYE,	180
BUFFLE HEAD DUCK,	184
LONG-TAILED DUCK,	188
HARLEQUIN DUCK,	195
SURF SCOTER,	201
AMERICAN SCOTER,	206
VELVET SCOTER,	210
WHITE-WINGED SCOTER,	212
STELLER'S DUCK,	216
SPECTACLED EIDER,	219
AMERICAN EIDER,	222
EIDER,	225
PACIFIC EIDER,	229
KING EIDER,	234
RUDDY DUCK,	237
MASKED DUCK,	242
AMERICAN MERGANSER,	245
RED-BREASTED MERGANSER,	249
HOODED MERGANSER,	254
SMEW,	259
APPENDIX,	263
L'ENVOI,	301
INDEX,	303

LIST OF ILLUSTRATIONS.

Portrait of the Author,		*Frontispiece*
1. Whistling Swan,		*Opposite page* 19
2. Trumpeter Swan,		" " 28
3. Whooping Swan,		" " 31
4. Blue Goose,		" " 33
5. Lesser Snow Goose,		" " 35
6. Greater Snow Goose,		" " 39
7. Ross's Snow Goose,		" " 43
8. White-Fronted Goose,		" " 45
9. Bean Goose,		" " 50
10. Emperor Goose,		" " 52
11. Canada Goose,		" " 57
12. Hutchins' Goose,		" " 69
13. White-Cheeked Goose,		" " 72
14. Cackling Goose,		" " 74
15. Barnacle Goose,		" " 78
16. Brant Goose,		" " 80
17. Black Brant,		" " 84
18. Wood Duck,		" " 87
19. Black-Bellied Tree Duck,		" " 92
20. Fulvous Tree Duck,		" " 95
21. Ruddy Sheldrake,		" " 97
22. Mallard,		" " 100
23. Dusky Duck,		" " 106
24. Florida Dusky Duck,		" " 109
25. Mottled Duck,		" " 111
26. Gadwall,		" " 113
27. European Widgeon,		" " 116
28. Widgeon,		" " 118
29. Sprigtail,		" " 122
30. Blue-Winged Teal,		" " 128
31. Cinnamon Teal,		" " 132
32. European Teal,		" " 134

LIST OF ILLUSTRATIONS.

33.	Green-Winged Teal,	Opposite page	136
34.	Shoveler,	" "	140
35.	Rufous-Crested Duck,	" "	144
36.	Canvas Back,	" "	147
37.	Red Head,	" "	154
38.	Scaup Duck,	" "	160
39.	Lesser Scaup Duck,	" "	164
40.	Ringed-Neck Duck,	" "	169
41.	Labrador Duck,	" "	172
42.	Golden Eye,	" "	176
43.	Barrow's Golden-Eye,	" "	180
44.	Buffle Head Duck,	" "	184
45.	Long-Tailed Duck, *Summer plumage*,	" "	188
46.	Long-Tailed Duck, *Winter plumage*,	" "	190
47.	Harlequin Duck,	" "	195
48.	Surf Scoter,	" "	201
49.	American Scoter,	" "	206
50.	Velvet Scoter,	" "	210
51.	White-Winged Scoter,	" "	212
52.	Steller's Duck,	" "	216
53.	Spectacled Eider,	" "	219
54.	American Eider,	" "	222
55.	Eider,	" "	225
56.	Pacific Eider,	" "	229
57.	King Eider,	" "	234
58.	Ruddy Duck,	" "	237
59.	Masked Duck,	" "	242
60.	American Merganser,	" "	245
61.	Red-Breasted Merganser,	" "	249
62.	Hooded Merganser,	" "	254
63.	Smew,	" "	259

INTRODUCTION.

THE family of the Anatidæ is composed of web-footed, swimming birds, having a bill covered with a soft skin, and a protuberance, sometimes hardly perceptible, at the tip, and contains the Swan, Geese, Ducks, and Mergansers, constituting Huxley's order CHENOMORPHÆ (Greek χήν, *chen*, a goose, + μορφή, *morphé*, form).

The family is divided into several subfamilies, the number varying according to the views an ornithologist may have as to their necessity, but never less than five, viz.: CYGNINÆ, Swan; ANSERINÆ, Geese; ANATINÆ, Fresh-Water Ducks; FULIGULINÆ, Sea Ducks, and MERGINÆ, Mergansers. In this book the subfamilies are seven, as, in addition to those just named, there have been adopted, PLECTROPTERINÆ, in which, among several other species all exotic, is included the genus Æx represented in North America by our beautiful Wood Duck (and which in most lists is placed far from its apparently true position), and ERISMATURINÆ, containing the spine- or stiff-tail ducks. In addition to these there are four other subfamilies; ANSERANATINÆ, CEREOPSINÆ, CHENONETTINÆ, and MERGANETTINÆ, whose species are all exotic to this continent.

These eleven subfamilies possess something like two hundred species, about sixty of which are found in North America. A conspicuous feature of these birds is a hard bony expansion at the end of the bill, occasionally occupying the whole tip and frequently bent over, forming a hook. This is called the nail, whence the Family is sometimes named UNGUIROSTRES. (Latin *Ungus*, nail, and *rostrum*, beak).

The ANATINÆ and the GALLINÆ are probably, to those who are not ornithologists, the most familiar of the feathered creatures. Like the Gallinaceous birds, the Water Fowl bear a very important relation to man, as they are the source of all domesticated races of web-footed birds, and they provide one of the chief means of subsistence to the inhabitants of boreal regions. Among civilized people they are regarded also as of great value from an economic point of view.

Usually these birds have a stout, full, rather heavy body, with a moderate or short neck (exceedingly long in the Swan), short legs, placed posteriorly in most instances, and generally hidden in the body feathers halfway to the heel, with the tarsus covered with scutellate or reticulate scales, sometimes with both, as in DENDROCYGNA; feet palmated, hind toe simple or lobed; oil gland present, and a large and fleshy tongue. Bill various in shape, from broad and flat, which is perhaps most usual, to long and very narrow. Lamellæ (plates or toothlike processes inside edge of bill), are frequently present, sometimes exceedingly prominent, numerous, and close together, and vary from those like the teeth of a fine comb suitable for sifting ooze, etc., to a rather coarse hooklike form, pointing backward to prevent the escape of any prey that may have been seized.

The sternum is broad and rather long, with a notch on each side, and sometimes the keel is hollowed out for the reception of the windpipe. This organ exhibits curious modifications in the various species. In some of the Swan it enters a hollow in the sternum, doubles on itself, forming a coil, and then emerges, passing onward to the lung. In certain species of Geese it forms a coil between the skin and breast muscles; and in a large number of the Ducks and Mergansers, several lower rings of the trachea are united together and enlarged, producing a capsule in the throat. These convoluted windpipes increase the volume of the voice, as in the case of the Trumpeter Swan, and in numbers of other Families the twisting and winding of this organ are carried to an extreme within the breast bone, as is seen in the Whooping Crane (*Grus americana*) and other species.

The wings vary in shape and in comparative length to the body. Some species have these very short, and they are moved with great rapidity, sometimes appearing devoid of outline so swift is their action, and their possessors go buzzing through the air more like insects than birds. Again the wings are long and pointed, and when the bird is flying are moved more slowly. Most of the Anatidæ, however, are rapid flyers, and even large species like Swan and Geese, although their flight may appear labored, proceed with much speed.

The plumage is dense and consists of a coating of down next to the skin, protected by the overlapping outer feathers, affording a very warm covering. Most of the species have a subdued coloring, but some are arrayed in a gorgeous dress of many hues, frequently exhibiting the brilliancy of metallic iridescence. The tail is of various shapes, rounded, cuneate, or with the median pair of feathers moderately or greatly elongated. The bills also vary greatly, from those that are broad, low, and flat, through a shape short, high at base, and rather pointed at tip, to one long, narrow, hooked, and serrated. The bill is covered by a skin, which in the Swan extends to the eye, leaving the lores bare. The sternum or breast bone being broad and flat with little or no keel, the pectoral muscles are consequently wide but not deep, differing in this respect from gallinaceous birds, which have a large keel to the sternum, and correspondingly deep breast muscles.

As I have already mentioned, the economic importance of the species of the ANATIDÆ is very great, and fortunately therefore their broods are large, and their numbers, although very much lessened in past years by constant slaughter, are fairly maintained in some portions of the continent. Of course, among so many kinds there is a great diversity in the quality of the flesh, and while some are eagerly sought for their high excellence there are others of which little can be said in praise. Those species that subsist upon rank grasses or animal substances are usually impregnated with the flavor of their food, and therefore not greatly desired for the table. Of these may be enumerated many of the Sea Ducks, some Geese, and the Mergansers. The birds of this Family place their nests (which are mostly formed of feathers and lined with down, plucked from the breast of the female), in the majority of cases upon the ground, but some build amid the branches of trees or occupy hollows in the trunk, and a few even seek holes in the banks, near streams. The eggs number from eight to twenty, are without markings, and vary in color from white to pale green. The young run and swim as soon as they escape from the shell, either seeking the water themselves, or else, as in the case of those hatched in a tree, are carried to it in the bill of the female. She incubates the eggs and cares for the young, in certain species the male assisting in watching over the brood; but generally the males are very remiss in these duties, and, especially among the Sea Ducks, frequently desert the females after incubation commences, and go away by themselves, forming a group of idle fellows, whose only idea of life is amusement and sustenance.

Between the sexes of the Ducks and Mergansers great difference in the color of the plumage is observable, males and females rarely resembling each other either in the hues or markings of their feathers, but among the Swan and Geese the sexes are similar. One characteristic mark of many species of Ducks is the speculum, or conspicuous spot on the wing formed by the coloration, often metallic, of the terminal portion of the secondaries. This sometimes serves to identify the species, especially in the case of the female, and is frequently of brilliant hues in both sexes, though brighter always in the male.

The various groups into which the Family has been divided are closely united, and although there are many artificial sections easily recognized among them, known as genera, yet all the species are more or less nearly related, and the Family is a very compact one, and easily distinguished from all others.

The Wild Fowl are migratory; some, indeed the great majority,—comprising all those breeding in boreal regions,—pass over an immense extent of the continent twice a year, spring and autumn. On such occasions they proceed in great flocks, usually some veteran bird leading the way, guided by the experience derived from travels of many years. The large species, Swan and Geese, journey in a V-shaped formation; Ducks also frequently adopt this same method, but they often also travel in a curved line, occasionally even all abreast. This last formation is not continued for any great distance. The few species inhabiting the temperate portions of North America, and which breed there, make very brief migrations, if indeed any at all. North America at one time probably contained more Wild Fowl than any other country of the globe, and even in the recollection of some living, the birds came down from the Northland during the autumn in numbers that were incredible, promising a continuance of the race forever. I have myself seen great masses of Ducks, and also of Geese, rise at one time from the water in so dense a cloud as to obscure the sky, and every suitable water-covered spot held some member of the Family throughout our limits. But those great armies of Wild Fowl will be seen no more in our land, only the survivors of their broken ranks. Let these, then, have the protection which is their due, and our advantage and profit to accord; stop all spring shooting within our borders, a time when the birds not only are usually poor in flesh, but are mated and journeying northward in obedience to the command, "be fruitful and multiply"; frown down all such barbarous customs as "killing for count," and then, with the impartial enforcement of the laws upon all the people, a remnant at least of our noble Water Fowl may be preserved to future generations.

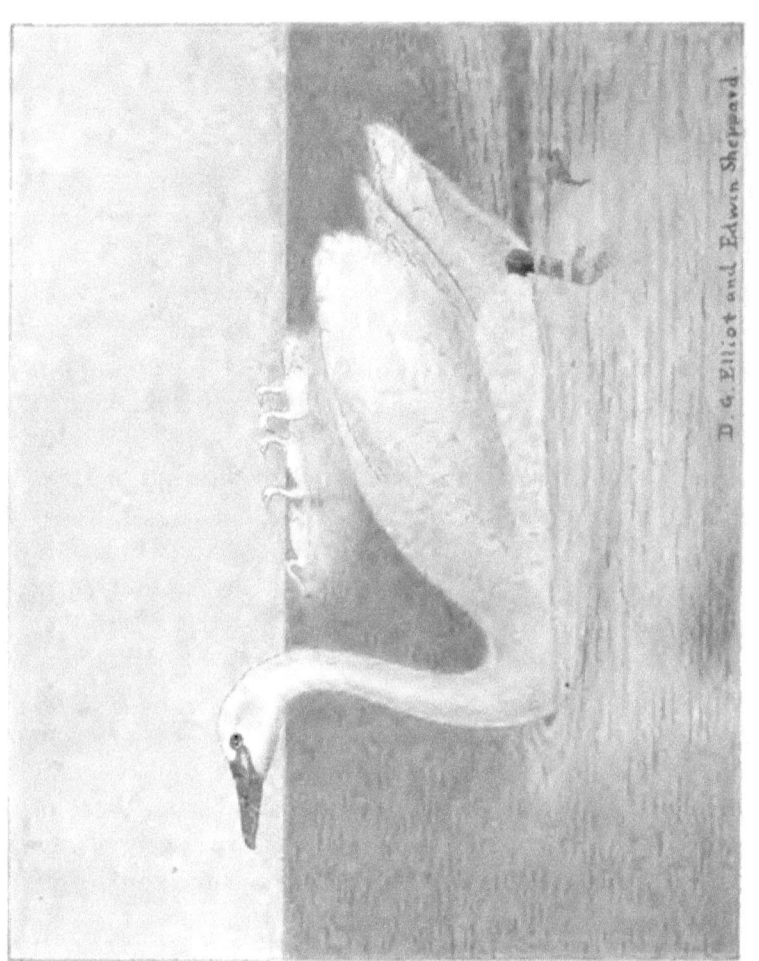

1. Whistling Swan.

WATER FOWL.

WHISTLING SWAN.

OF the two species of Swan indigenous to North America, the present one is the smaller and more widely dispersed. It ranges in the northern portions of the continent from the Atlantic to the Pacific, and from the Arctic regions south to California on the west, and to the Carolinas on the east coast, being very abundant in winter in Currituck Sound, North Carolina. It is also found in the Mississippi Valley south to the Gulf of Mexico, and is common in Galveston Bay, Texas. It breeds on both sides of the mountains in the Arctic regions; in the interior chiefly, if it stops short of the Arctic Ocean, but if not, then on the coast and contiguous islands of that sea. It nests in the marshes at the mouth of the Yukon, and also along that great river above the Delta, and on the shores about St. Michael's. On the Alaskan coast by the Arctic Sea this Swan is rare, and it is not found in any of the islands, nor on the Siberian shore of Behring Sea, but is met with on the far western islands of the Aleutian chain, though it does not breed on any of them.

This species arrives near the mouth of the Yukon the latter part of April or beginning of May, coming down the river from the interior, and not along the coast from the south, and as they return the same way, it is supposed they cross the mountains near the head waters of this stream. The nest is placed upon an island in some small lake, or on its borders. It is a large structure—sometimes six feet long, four and a half wide, and two high,—composed of grass, dead leaves, moss, and other rubbish. The eggs are pure white or fulvous, and the number seems to vary from one to six, but I should imagine the latter to be very exceptional, or else there must be a great mortality among the cygnets, as it is unusual in winter to see a pair of these birds accompanied by more than two young. The eggs usually lie hidden in the moss, artfully concealed by the female. By the last of June the young are hatched, and are led by the parents to the nearest water, and soon after the adults moult, when many are killed by the natives, who spear the defenseless birds unable to fly, and sometimes capture them alive. Toward the last of September they gather in flocks, and by the second week in October all have departed for southern waters.

While on their journey to and from their winter quarters, this Swan deserts the coasts and proceeds inland, traveling at a great height and making long flights without halting. The migrating host from the far north, on entering the United States, separates into three divisions: the western keeping to the Pacific slopes, the center to the valley of the Mississippi (where the species is much more rare than the Trumpeter Swan), and the remainder, or eastern flank, bearing away to the broad waters of the Chesapeake and the sounds of North Carolina. The flocks are strung out in long, divergent lines, headed by some sagacious old bird, whose powerful wings beat the air, and break a passage, so to speak, for those that follow. Whenever he becomes fatigued by this extra labor, he utters a note that seems to be well understood by the others, and falling out of line, his place is supplied by another; the late leader taking a position back in the ranks. Before alighting, the ground and water beneath them are carefully examined for any hidden foe, and after the leader is satisfied that all is right, with graceful curves, and easy sailings on their great wings, the birds alight upon the water and commence to feed.

This Swan makes its appearance on the Atlantic coast about the beginning of November. It is rare north of the Chesapeake, but very numerous on the littoral waters of North Carolina, and apparently is more abundant there every year. They arrive in small flocks, succeeding each other on some days in rapid succession; passing at times over the beach, again over the ocean, or the water inside the beach. They fly usually at a considerable height, and the beat of their great wings is so short as to give these the appearance of being almost motionless. The black feet extend beyond the tail, and with the long neck stretched out to its fullest extent, the great birds survey the landscape beneath them. Occasionally the peculiar flageolet-like note is uttered by the leader, the syllables sounding something like *Whō, whŏ-whŏ*, in a very high key, and this being responded to by other members of the flock, a chorus of weird sounds from out the upper air floats downward to the ear of the watcher below. Beautiful indeed, the splendid birds appear, sailing onward in the blue sky, the bright rays of a midday sun glancing from their immaculate plumage, causing it to glisten with the sheen of burnished silver, or, if the birds are passing directly overhead, the light streaming through the feathers of the wings reflects on the under side and also on the body, a glow like the faint blush on the petals of a rose. With redoubled cries the glad birds welcome the well-known waters of their winter home, and gradually lowering themselves from their lofty altitudes, turn head to wind, and checking their momentum by a few rapid wing beats, launch themselves into the waters of the sound. Should there be any Swan in the vicinity, and the newcomers are the fewest in number, they swim to them, otherwise little attention is paid to other flocks.

Their journey having provided them with sharpened appetites, they soon commence to feed by immersing their heads and dragging up the grass from the bottom. If the water is deeper than the length of the neck, the hinder part of the body is tilted up and held in position by paddling with the feet, until a quantity of tender grass is torn from the bottom. While feeding, usually one or more birds keep a lookout for approaching danger, and should any be descried, a warning note is sounded, and the flock begins to swim away, heading to windward, if possible. If undisturbed, Swan are very noisy, keeping up a continual medley of cries, usually uttered in so high a key as to render it impossible to imitate without artificial aid, but if alarmed, the birds immediately become silent, and remain so until the object of their fear has departed. Upon the water this Swan floats lightly and presents a beautiful appearance. When congregated together in large numbers they seem, from a distance, like snow islands, so pure and white is their plumage. As they move gracefully along, propelled by a powerful shove of one webbed foot after another, the neck is usually carried upright, though occasionally with a graceful curve the head is lowered for a sip of water, or to seize upon a morsel of floating grass.

Where Swan have been feeding for any length of time great holes are hollowed in the bottom, the mud or sand having been scooped out by their powerful bills and feet and piled up on the side, and when the water is moderately shallow, I have known a sailboat to be frequently grounded upon the lumps thus formed. In this way these birds do great damage to feeding grounds, and destroy very much more edible grass than they consume. For this reason they are not altogether regarded with favor by sportsmen, as they soon render useless large tracts of grass-covered bottom to which Ducks and Geese would resort for a long time, but which they are forced to desert on account of the wasteful destruction of their food committed by the Swan. As a rule this species pays but little attention to decoys, or wooden representatives placed among a number of live Geese tied out for the same purpose. Most of the birds that are procured are shot from points over which the Swan fly as they pass up and down their feeding grounds; or are killed from boats sailed down upon them before the wind. Swan being so large and heavy cannot easily take wing, but are obliged to force themselves over the water, and against the wind, by rapid and powerful beats of the wings and feet, until, obtaining the requisite momentum, they are lifted into the air. Of course then, when a boat approaches them down wind, they are obliged as it were, to run toward it, before they are able to fly away, and it not unfrequently happens that a person in a sailboat can thus get within shooting distance of these wary birds. Large shot and heavy loads of powder are needed to bring them down; an ounce or an ounce and a half of double T., with five drams of powder, is a good load for them. When a flock is shot on the wing the birds rarely swerve from their course, and even when one falls the rest close up the gap and keep on as if nothing had happened. If very near the sportsman, however, when he fires, the birds will swing to one side or the other, but immediately after return and continue on their original direct route. If they see anything unusual in their line of flight the leader immediately slightly alters his course, closely followed in regular order by the birds that succeed him. When a Swan is killed in the air, he doubles all up in falling; head neck, wings, and legs appearing to be mixed up together; and on striking the water, unless this is very deep, the weight of the bird and the impetus acquired by its fall will frequently carry it quite to the bottom. I have known them to fall where the water was fully three feet deep, and rise to the surface covered with mud obtained from the bottom. When mortally wounded in the air, the Swan will usually set its wings and sail slowly toward the earth or water, whichever it may happen to reach.

The song of the dying Swan has been the theme of poets for centuries and is generally considered one of those pleasing myths that are handed down through the ages. I had killed many Swan and never heard aught from them at any time, save the familiar notes that reach the ears of everyone in their vicinity. But once, when shooting in Currituck Sound over water belonging to a club of which I am a member, in company with a friend, Mr. F. W. Leggett of New York, a number of Swan passed over us at a considerable height. We fired at them, and one splendid bird was mortally hurt. On receiving his wound the wings became fixed and he commenced at once his song, which was continued until the water was reached, nearly half a mile away. I am perfectly familiar with every note a Swan is accustomed to utter, but never before nor since have I heard any like those sung by this stricken bird. Most plaintive in character and musical in tone, it sounded at times like the soft running of the notes in an octave.

> " And now 'twas like all instruments,
> Now like a lonely flute;
> And now it is an angel's song
> Which makes the heavens be mute,"

and as the sound was borne to us, mellowed by the distance, we stood astonished, and could only exclaim, " We have heard the song of the dying Swan."

I made inquiries among the gunners as to whether any of them had ever heard notes different from those usually uttered by the Swan, when one was mortally wounded, and some said they had, and on my asking them what kind they were, they described something similar to those we had heard and of which I have endeavored to give an idea. We recovered the bird, which was an adult in perfect plumage, and the skin made into a screen adorns the drawing room of my friend.

The young of this species is gray, sometimes lead color during its first year, and the bill is soft and reddish in hue. In the second year the plumage is lighter, and the bill white, becoming black in the third year, when the plumage, though white, is mottled with gray; the head and neck especially showing but little white. It is probable that it takes fully five years before the pure white dress is assumed and the bird becomes such an ornamental object. The flesh of the old birds is tough and unfit to eat, and boiling is necessary before it can be masticated, but the young or cygnets are tender and well flavored. The Swan is supposed to live to a great age, but this is one of those problems very difficult to solve. The length of time the domesticated bird may live is no criterion (on account of its altered mode of life) to estimate the age of the wild Swan, and of course for the latter it is impossible to acquire any data to enable a judgment to be formed. From fifteen to twenty years,

I should suppose would be the average limit of the bird's existence.

This species loves to keep near the shores of marshes and islands, and is frequently seen standing on the bank dressing its feathers. This habit is taken advantage of by the gunner, who selects a day when the wind is blowing hard, and landing upon the opposite side of the marsh or island on which the birds are standing, and availing himself of the shelter of the reeds, creeps upon the unsuspecting Swan, who cannot hear him on account of the wind, and shoots them down at close quarters. When the weather becomes severe and the sounds and bays are frozen, the Swan are seen standing on the ice, surrounded by the more watchful geese. If the severe weather continues to close the waters, the birds depart for more southern climes, until a change of temperature occurs, when they at once return to their old quarters.

At the advent of spring the Swan begin to show signs of uneasiness, and to make preparations for their long journey to the northward. They gather in large flocks and pass much of the time preening their feathers, keeping up a constant flow of loud notes, as though discussing the period of their departure and the method and direction of their course. At length all being in readiness, with loud screams and many *Whŏ-whŏ's* they mount into the air, and in long lines wing their way toward their breeding places amid the frozen north. It has been estimated that Swan travel at the rate of one hundred miles an hour with a moderate wind in their favor to help them along. The American Swan is monogamous, and once mated the pair are presumed to be faithful for life. The young keep with their parents for the first year, and these little families are only parted during that period by the death of its members. A wounded Swan is very difficult to capture, for it immediately swims away right in the wind's "eye," and so rapidly can it propel itself by its broad feet that a man in a boat has great difficulty in capturing it. When overtaken, it is found to be no mean antagonist, for it can deal severe blows with its wings, sufficiently powerful at times to break a man's arm, while the great feet are capable of committing severe injury with their long claws. It is therefore necessary to be somewhat careful in approaching a wounded Swan.

In addition to its smaller size the present species can be distinguished from the Trumpeter Swan by the presence of a yellow oblong spot on the naked skin near the eye, this part in the other species being all black. It weighs from twelve to twenty pounds, some exceptionally large birds perhaps a few pounds more. In Louisiana this species is called Cygne.

CYGNUS COLUMBIANUS.

Geographical Distribution.—America, generally; Commander Islands, Kamchatka. Accidental in Scotland. Breeds in Arctic regions.

Adult.—Plumage, pure white; occasionally individuals have rust color spots or blotches on head and neck, sometimes also on the body. Lores naked, with a small yellow spot. Bill and feet black. Total length, about 50 to 55 inches; wing average, 21; tarsus, 4½; culmen, 4.

Young.—General color, gray; sometimes nearly a lead color during the first year, and the bill reddish in hue. Second year the plumage is lighter and the bill is white, turning to black in the third year, when the plumage is white, mottled with gray on the body, the head and neck being mostly all gray. It requires about five years before the plumage becomes entirely pure white.

Downy Young.—Pure white. Bill, legs, and feet, yellow. From a specimen taken at Franklin Bay, Arctic America, by MacFarlane in 1869, now in the Philadelphia Academy of Natural Sciences.

TRUMPETER SWAN.

THIS splendid bird differs from the American or Whistling Swan in its larger size, absence of yellow near the eye, and the peculiar arrangement of the windpipe. It is found in the interior of North America and on the Pacific coast, but never appears on the shores of the Atlantic unless as a straggler. It breeds on the islands and in the low reedy grounds around Hudson Bay, also in the Barren Grounds near the Arctic coast, and in the interior probably on both sides of the mountains, but is not known to breed in Alaska. A single specimen was procured by Dall at Fort Yukon, which is the only record given of its appearance in the Territory. In the United States, the Trumpeter, in the interior, winters from Illinois to the Gulf of Mexico, and breeds from Iowa and Minnesota northward. The nest of this species is a large structure composed of grass, leaves, down, and feathers, and is placed usually on elevated ground. The eggs, which are a uniform chalky white with a granulated surface, are quite large, from four to four and a half inches long, and two and a half to three in breadth. From five to seven is the complement, and the young are hatched in July, and are led by the parents to the fresh-water ponds and lakes in the vicinity. In August the adults moult and are then for a time unable to fly, and about the beginning of September the birds commence to journey southward, and are among the first of the

2. Trumpeter Swan.

great migratory host to enter our limits, and also to leave them again in the spring.

The Trumpeter swims rapidly and easily, and when going before the wind raises its wings and uses them as sails to help itself along. It flies very high and in lengthened lines, like the Whistling Swan, and its speed in the air is about the same, possibly one hundred miles an hour under favorable conditions. Its voice is very different from that of the other species, being loud and sonorous, resembling the notes of a French horn, the tone being caused by the various convolutions of the windpipe.

I do not think that this species, in the localities it frequents, is as numerous as is the Whistling Swan in its habitats. It is the prevailing species in California, where it visits the inland fresh waters, and is apparently most abundant on the rivers emptying into the lower Mississippi, along the Gulf of Mexico, and in Western Texas, where it is fairly common in winter. It does not differ in its habits from the other species to any appreciable extent. It feeds on roots of aquatic plants, grasses, shell fish, crustacea, etc., and procures its food in the same way as the Whistling Swan by immersing the head and neck, and pulling the desired objects from the bottom. It associates in small flocks by itself and is very shy and suspicious. The weight of this Swan varies from twenty to thirty pounds, being, on the average, considerably heavier than the other species. It is a trim, well-shaped, handsome bird, and when congregated in numbers on the water has all the beautiful appearance characteristic of its relative.

Cygne is the popular name given to this species in Louisiana, the same as that applied to the Whistling Swan.

CYGNUS BUCCINATOR.

Geographical Distribution.—Interior of North America, west to the Pacific coast, from the Arctic regions to the Gulf of Mexico. Breeding from Northern United States, as Iowa and the Dakotas, northward. Accidental on the Atlantic coast.

Adult.—Entire plumage, white; sometimes a wash of rust color on the head. Bill, lores, and feet, black. Average total length, about 63 inches; wing, $24\frac{1}{4}$; tarsus, $4\frac{3}{4}$; culmen, $4\frac{1}{2}$.

Young.—General plumage, gray, with rust color on head and neck. Bill, basal end flesh color, dusky for remaining portion. Legs and feet, grayish.

3. Whooping Swan.

WHOOPING SWAN.

IT can hardly be considered that this Swan is a North American species, as it has never yet been found upon this Continent. Its claim to be included in our avi-fauna is based on the supposition that it is still a visitor to Greenland. The Whooping Swan is a native of the Old World, found throughout the British Islands and the Continent of Europe, going as far south in winter as Egypt and eastward through Asia to Japan.

It breeds in high northern latitudes in Iceland and Finnish Lapland, and in the vast marshes of the Arctic regions. The nest, which is very large, and said to be occupied by the same bird for a number of years should it survive, is placed on some tussock, and is composed of rushes, grass, and similar materials. Incubation lasts forty-two days, and the number of eggs, which are yellowish white, varies from four to seven, the former being the most usual. The young, which are generally hatched in June, are not able to fly until August, and are carefully guarded by the parents, who protect them from their numerous enemies, becoming the aggressors on slight provocation, and are antagonists not to be despised. It is a handsome bird, though, on account of its shorter neck, not so graceful as the Mute Swan (*Cygnus olor*), so commonly seen on ornamental waters in Europe. It frequently comes upon the land to pull up the grass, which it does in the manner of geese, and it walks easily if not gracefully.

The Wild Whooper is a very shy bird, and permits nothing of which it is suspicious to approach. It goes in moderate-sized flocks and the birds fly in V-shaped lines, and continually utter their trumpet call. In winter they gather together in considerable numbers. This Swan is a large bird and will weigh from twelve to twenty pounds. Although of greater dimensions, it bears more resemblance to Bewick's Swan than to any other European species, but is readily distinguished by having nearly two-thirds of the maxilla, or upper part of the bill, yellow. Swans mate for life, and the same pair will usually return to the last year's nest. Among young males, or old males which have lost their mates, fierce fights take place during the breeding season, or until most of them have become mated. The habits of the Whooping Swan are similar to those of the Mute Swan, which are known to all who have watched this bird in a domesticated state in Europe.

In Greenland this present species formerly used to breed, as stated by the Eskimo, near Godthaab, but was exterminated when moulting and unable to escape. It has occasionally reappeared in South Greenland during the past thirty or forty years, but so irregularly, and usually single individuals only, that it would seem these were merely stragglers coming from Iceland, where the bird is known to breed on the large marshes.

CYGNUS CYGNUS.

Geographical Distribution.—Northern parts of eastern hemisphere, occasional in Southern Greenland.

Adult.—Plumage, entirely white. Basal portion of bill and lores, yellow, this color surrounding the nostrils, remainder black. Legs and feet, black. Average total length about 57 inches; wing, 24; tarsus, 4; culmen, 4½.

Young.—General color, grayish brown. Bill, base and lores, greenish white; remainder black, with a reddish orange band across the nostrils.

Downy Young.—All white.

4. Blue Goose.

BLUE GOOSE.

FOR a long time this fine species was considered to be merely the young of the Snow Goose, although in its adult dress it bears no resemblance to that bird. Very little is known of either its economy or habits, and it is seldom seen upon any of our seacoasts, keeping chiefly to the Mississippi Valley, where it is a migrant, going in winter to the Gulf. The breeding grounds of this Goose are unknown, but the Eskimo say they are to be found in the interior of Labrador, among the impenetrable bogs and swamps that are so numerous in that country. It is refreshing to learn that some birds have inaccessible retreats where they can rear their young without molestation. According to Mr. G. Barnstone, this species crosses James Bay (in the southern part of Hudson Bay), coming from the eastern coast, while the Snow Goose comes down from the north, seeming evidently to indicate that their breeding places are distinct. Hearne, who met with this Goose in the last century, states that its flesh was very palatable, quite as good as the Snow Goose, and that it was seldom seen north of Churchill River, but very common at Fort York, and at Fort Albany. It is occasionally seen in company with the Snow Goose. The Blue Goose has been taken on the coast of Maine and at Grand Menan, but is very rare along the Atlantic. In the west it is more common and numbers are killed every winter, but it has not been found anywhere upon the shores of the Pacific. This species is usually distinguished from the Snow

Goose, as the Blue, or Blue Snow Goose, Bald-Headed Goose, White-Headed Goose, Oie Bleu, and Blue Brant in Louisiana, and in the north where all Snow Geese are called Waveys, as the Blue Wavey. It is a very handsome bird in its adult summer dress, the handsomest in my opinion of all our Geese, and doubtless could be domesticated and become an ornament to our farmyards.

CHEN CÆRULESCENS.

Geographical Distribution.—Hudson Bay, through interior of North America, along the valley of the Mississippi to the Gulf of Mexico. Very rare on the coast of Maine, but not found farther south on the shores of the Atlantic, nor anywhere on the Pacific.

Adult.—Head and upper part of neck, white; sometimes a blackish brown line extends from top of head along middle of hind neck. Rest of neck, breast, back, and wings, grayish brown. Wing coverts, and rump, bluish gray. Secondaries, blackish brown, edged with white. Primaries, blackish brown. Flanks, grayish brown; feathers, tipped with pale brown. Under parts, white or whitish; upper and under tail coverts, whitish. Tail, brownish gray, edged with white. Bill, pale pinkish; nail, white; a black line along the edges of the maxilla and mandible. Legs and feet, reddish color. Total length, about 28 inches; wing, 16; tarsus, $3\frac{1}{16}$; culmen, $2\frac{2}{10}$.

Young.—Like the adult, but with the head and neck dark grayish brown; chin only white.

5. Lesser Snow Goose.

LESSER SNOW GOOSE.

THIS bird, the smaller of the two Snow Geese, is the western representative form, ranging from the Arctic Sea, south in winter to the Gulf of Mexico and Southern California. It does not breed south of the Arctic circle, and during its migrations makes no tarrying in Alaska, stopping but a brief period to rest and feed on the marshes, and then continuing its journey northward. The flocks arrive on the Yukon from the beginning to the middle of May, and are more numerous in spring than in the autumn, when they return re-enforced in numbers by their young families. None pass the winter in any part of Alaska, and the species does not seem to visit the Aleutian Islands at any time. On their return journey when they appear in the United States, about the beginning of September, they come in flocks numbering sometimes over one hundred individuals, but are not seen upon the coast, performing their migrations apparently over the land.

In Washington and Oregon and throughout California this Snow Goose is very common in winter, and frequents the plains and marshes near the sea. It arrives in October, and remains until March, and like the larger species is shy and watchful. In the interior of the Continent and along the Mississippi Valley it is a regular migrant, and is quite abundant. It arrives there about the same time as the members of the western army do on the Pacific coast, from the beginning to the middle of October, flying very high in a long, extended curved line, not nearly so angular as the V-shaped ranks of the Canada and other Geese. With their snowy forms moving steadily along in the calm air, the outstretched wings tipped with black, glowing in the sun's rays with the faint blush of the rose, they present a most beautiful sight. Usually they fly silently with hardly a perceptible movement of the pinions, high above

> ". . . the landscape lying so far below
> With its towns and rivers and desert places,
> And the splendor of light above, and the glow
> Of the limitless blue ethereal spaces."

Occasionally, however, a solitary note like a softened *Honk* is borne from out the sky to the ear of the watcher beneath. Should they perceive a place that attracts them they begin to lower, at first gradually, sailing along on motionless wings until near the desired spot, and then descend rapidly in zigzag lines until the ground or water is almost reached, when with a few quick flaps they gently alight. It is difficult to get close to them, as they are very watchful, and if they become suspicious an alarm is sounded and the flock betakes itself to some other locality. Sometimes, in passing from one place to another, they fly low enough to give the concealed gunner a chance for a successful shot, but I have never known them to decoy at all well, and the majority of those procured are birds passing to and from their feeding grounds.

As an article of food I have never held this bird in any great esteem, for if it was tender it had very little flavor, and if the latter was clearly perceptible it was generally of that kind one would prefer to have absent. When this Goose first arrives it is very apt to be lean, having had but little time on its long journey to stop and feed sufficiently to fatten, but after a short stay upon the plains and waters of more southern climes, where food is abundant and easily obtained, it soon recuperates and becomes fat and in fine condition.

At times this species assembles in such multitudes as to give the landscape the appearance of being covered with snow, but if the sportsman, misled by their numbers, thinks he certainly can secure some individuals out of such a vast concourse, and attempts to get within shooting distance by any ordinary means, he will probably find himself greatly mistaken, for long before the desired spot is gained, he will see the vast white sheet rise, and countless wings winnow the air. Sometimes they will permit a wagon to be driven almost into their midst, or a man on horseback can charge at full speed and get up to them, and many are occasionally taken by these methods, but they soon learn what dangers to avoid, and are very successful in doing so, although they may immediately afterward be deceived by some more simple but novel stratagem. The young are always unsuspicious, and can easily be distinguished from the old birds, even in the air, by their grayish plumage, which makes them very noticeable among the pure white members of the flock, and at a little distance, they appear as if they had soiled their feathers in mud, which had afterward become dry.

The Lesser Snow Goose does not differ in appearance from the larger species, and it will be often necessary to measure a specimen to know to which form it belongs. Size is at all times a most unsatisfactory distinction. This bird is called Baily (white) Goose, by the Russians, and Oie Blanche and White Brant in Louisiana, and the same names are also applied to the succeeding form in that State.

CHEN HYPERBOREUS.

Geographical Distribution.—Western North America from the Valley of the Mississippi to the Pacific coast, and from Alaska to Southern California. Breeds within the Arctic circle.

Adult.—Primaries, black; their bases and coverts, ashy. Entire rest of plumage, white. Bill, purplish red; nail, white; space between maxilla and mandible, black. Legs and feet, orange red. Iris, dark brown. Total length, about $25\frac{1}{2}$ inches; wing, $15\frac{3}{4}$; tarsus, 4; culmen, $2\frac{1}{10}$.

Young.—Head, neck, and upper parts, light gray; feathers of back, tertials, and wing coverts, with dark centers, and edged with white. Primaries, black. Rest of plumage, white.

6. Greater Snow Goose.

GREATER SNOW GOOSE.

IT is somewhat difficult to define accurately the limits of the present bird and the preceding, when there is nothing to distinguish them from each other but a difference of a few inches in their total lengths; and unfortunately wild birds object to be measured, so it is impossible to verify one's observations with that degree of certainty so much desired by all naturalists, and so rarely obtained. But since it has been decided that there are two forms of this Snow Goose in North America, the present is considered as that one which is found east of the Mississippi Valley and chiefly along the coast of the Atlantic Ocean, going occasionally as far south as Cuba. Like its smaller relative its breeding places are in the far north, on the Barren Grounds, and on the borders of the Arctic Ocean east of the Mackenzie River. It is very common in summer during its migrations about Hudson Bay, so abundant that formerly a single hunter has been known to kill a thousand to twelve hundred in a season. A much smaller number than this has to suffice at the present time. Snow Geese flock by themselves, and although they may be feeding on the same marsh or plain, or stretch of water with other Geese, never mingle with them. They feed chiefly on grass which, if on land, they bite off with the side motion of the head and jerk of the neck in precisely the same way as tame Geese are wont to do. These birds also eat bulbous roots and soft portions of various water plants, and their peculiarly shaped bills are admirably adapted for cutting or pulling apart such kind of food. In summer, according to Richardson, in the northern regions they feed on berries, and frequent the shores of lakes and rivers, and seldom are seen on the water except at night or when moulting. MacFarlane discovered on an island, in a lake near Liverpool Bay, some nests of the Snow Goose which were mere holes or depressions in the sandy soil well lined with down. The eggs are large and yellowish-white. The young are on the wing by the middle of August, and feed at first chiefly on insects and rushes, and later on berries. They are excellent for the table, and form, with the adults, the staple article of food for the natives in that region.

Previous to starting on their southern journey the birds desert the marshes, and keep near the edge of the water as it ebbs and flows, dressing their feathers continually. Then, all being ready, they take advantage of the first wind from the north and, mounting into the air, are borne at a high speed by their own efforts and favoring breezes, away from the ice-bound shores to sunnier climes, leaving the cheerless land that had been their summer home to lapse into the silence and darkness of a continued night.

> " With mingled sounds of horns and bells
> A far-heard clang, the Wild Geese fly,
> Storm-sent from Arctic moors and fells,
> Like a great arrow through the sky."

On the northern portion of the Atlantic coast the Snow Goose cannot be said to be common, and in many parts is seldom seen. Small flocks are occasionally met with on the waters of Long Island, but the species becomes more abundant on the shores of New Jersey and the coasts of Virginia and North Carolina, where, in the latter State in the vicinity of Cape

Hatteras, and along the beaches and inlets of Albemarle Sound, it sometimes congregates in great multitudes. Occasionally flocks of considerable size may be seen on the inner beach of Currituck Sound where the water is brackish, but the birds do not remain any length of time in such situations. They present a beautiful sight as they stand in long lines upon the beach, their pure, immaculate plumage shining like snow in the sun, against the black mud of the marshes or the dingy hues of the shore. It is very difficult to approach them at such times, as they are exceedingly watchful and wary, but occasionally a few may leave the main body, and, if flying by, will draw perhaps sufficiently near to Geese decoys, or live Geese tied out in front of a blind, to afford an opportunity for a shot. The chances are better, however, for the sportsman, when these Geese are moving in small flocks of six or seven, as they are then more apt to come near the shore looking for favorable feeding places, or spots on the beach to sand themselves.

It is a very silent species, and save for exceptional reasons such as becoming alarmed, or when about to migrate, it rarely utters a sound. The bill of this Goose is very strong and highly colored, with the edges of the upper and lower parts widely gaping, giving it a grinning expression, but it is an instrument admirably adapted for the employment given it by the owner, that of forcibly pulling reeds, grasses, etc., up by the roots. Beside the name of Snow Goose, both this species and the allied form are known throughout the land as White Brant. In the "Fur countries" the Greater Snow Goose is called the Common Wavey, also along the Atlantic coast it is known as Red Goose, probably from the color of its bill and legs, and Texas Goose, for no reason that I can see whatever.

CHEN HYPERBOREUS NIVALIS.

Geographical Distribution.—Shores of the Arctic Ocean east of the Mackenzie River, occasionally going south as far as Cuba, and from the Valley of the Mississippi to the Atlantic Ocean. Breeds in the Arctic regions.

Adult.—Resembles the Lesser Snow Goose in the color of the plumage, but is somewhat larger in its measurements. Average total length, 34 inches; wing, $17\frac{4}{10}$; tarsus, $3\frac{8}{10}$; culmen, $2\frac{6}{10}$. The average difference between the Greater and Lesser Snow Geese as given in Ridgway's "Manual" is, total length, 9 inches; wing, $1\frac{7}{10}$; culmen, $\frac{1}{2}$; tarsus, $\frac{7}{10}$.

From these measurements it will be perceived that it would be practically hopeless to try to originate any method for accurately separating these birds, for a specimen of the Lesser Snow Goose might be found larger than one of its supposed big "brothers."

Downy Young.—Lores, dusky. Two black stripes from bill, passing above and beneath the eye. Top of head, dark olive brown. Sides of head, neck, and entire under parts, light yellow. Upper parts, dark olive brown. Bill, black; nail, yellowish white. Specimen in Academy of Natural Sciences, procured 10th July, 1893, at Glacier Valley, Greenland, together with the adult female; Lieutenant Peary's Expedition.

7. Ross's Snow Goose.

ROSS'S SNOW GOOSE.

THIS is one of the smallest Geese known, a fully adult bird weighing only about two and a half to three pounds. It is remarkable for the curious carunculations at the base of the bill. It breeds in some part of the Arctic regions, but its nest and eggs have not as yet been discovered. Ross's Goose has never been found on the Atlantic coast of the United States, but it is not uncommon in parts of California in winter, and has been seen in the San Joaquin Valley in considerable numbers. Its journey to the south seems to lie to the westward entirely, and but little is known of its habits beyond the few observations made in California, and I have always regarded it as the rarest of our Geese. It has a cry like that of the Cackling Goose, and usually associates with the Lesser Snow Goose, and accompanies flocks of that bird in the air, flying on one side or the other, or else is scattered throughout the ranks of the main body of the larger birds. It was discovered by Hearne, who called it the Horned Wavey and said that two or three hundred miles west of Churchill, which is near the west shore of Hudson Bay, he saw them in as large flocks as the Common Wavey or Snow Goose. The flesh, he says, was extremely delicate, and as a proof of it he ate two of them one night for supper, which was doing very well, even for an Arctic appetite. It is a beautiful little bird, and it is to be regretted that more do not enter within our limits.

EXANTHEMOPS ROSSII.

Geographical Distribution.—Arctic America, south in winter to Southern California, east to Montana.

Adult.—Entire plumage, pure white, with the exception of the primaries, which are black. Bill, dull red; nail, white, without any black line along the gape. Basal portion of maxilla covered with wart-like excrescences. Legs and feet, reddish. Average total length, 23 inches; wing, $14\frac{3}{10}$; tarsus, $2\frac{6}{10}$; culmen, $1\frac{6}{10}$.

Young.—Resemble those of the Lesser Snow Goose, but are of a generally lighter color.

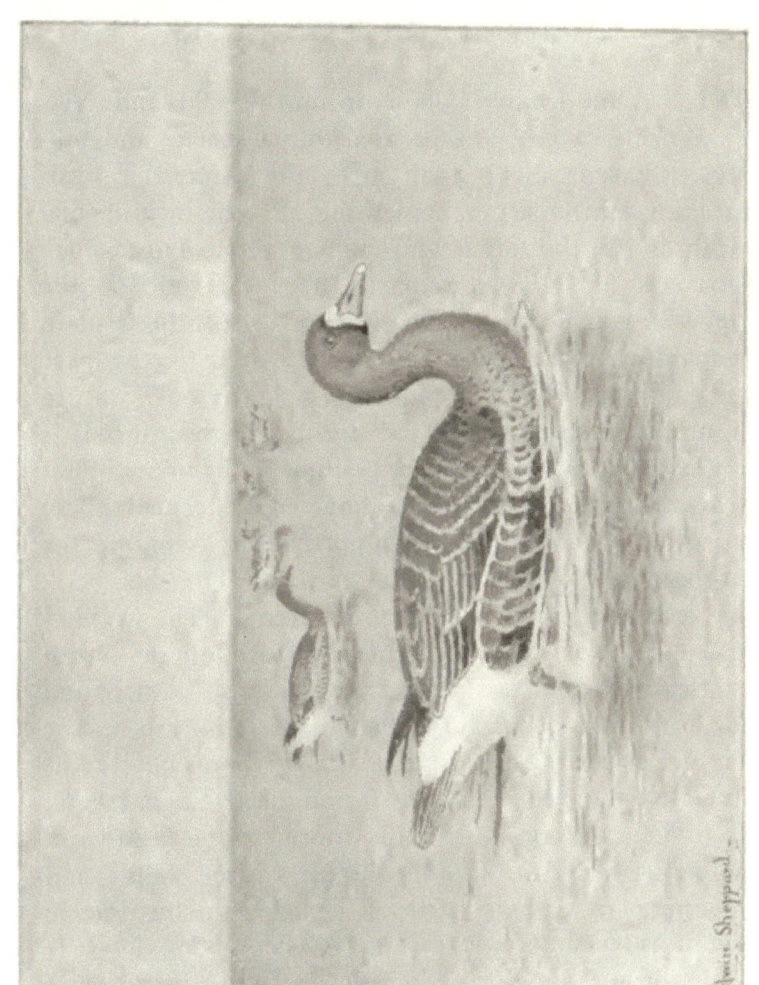

8. White-Fronted Goose.

WHITE-FRONTED GOOSE.

THE White-fronted Geese from the Old and New Worlds have been separated into a species and subspecies, based solely upon size; the American birds averaging a little larger, something like one inch in total length and in the tarsus and culmen about half an inch each. As all critical remarks are reserved for the Appendix it is not necessary here to discuss the wisdom of separating these birds, but merely to state that as there is no difference in their plumage, and the only way to distinguish a specimen (if two forms are recognized) is by the locality and the tape-line (and one cannot always then be certain), I have not deemed these distinctions as of sufficient importance to separate the European and American examples, and in this book have considered them as one species. The White-fronted Goose is found generally throughout North America from the Arctic Sea to the Gulf of Mexico, and Cuba, and also occurs in Greenland. It is rare on the Atlantic coast of the United States, occasional individuals having been taken as far south as Long Island, but in its migrations it tends more to the westward, is found in winter throughout the Mississippi Valley, and is common in various parts of Texas. On the Pacific coast it is very abundant from Alaska to Mexico. It breeds throughout the Arctic regions from the Atlantic to the Pacific, nesting on the lower Anderson River from its mouth to Fort Yukon; frequents the Siberian shore of Behring Straits, is found on the Commander Islands, and various others in Behring Sea, and also about the islands of the Aleutian chain, but is not known to breed on any of the last named. At St. Michael Island this species is abundant in May, and is called the Tundrina Goose or Low-ground Goose. Mr. MacFarlane, who found many of their nests on the Anderson River, states that these were depressions in the soil, and in nearly every instance lined with dried grass, down, and feathers. In Alaska the nest is placed in a hollow in the sand, or on the bank of some large pond or grassy flat, and is lined, like those on the eastern side of the continent, with grass or moss. But as the eggs are laid, the female plucks down from her breast, increasing the quantity until, the complement having been reached, the eggs are fairly covered. These are dull white, very similar to those of the Snow Goose.

The White-fronted Goose reaches its breeding grounds early in May, and is a very noisy bird, and announces its presence by loud cries. Mating accomplished they scatter in pairs, selecting sites for their nests and preparing for the serious duties of incubation. They remain about the fresh-water lakes and ponds, and subsist upon grasses, berries, and such like food. The parents attend the young until the latter are able to fly, usually in August, and later gather together in large flocks preparatory to starting on their southern journey, which is begun toward the last of September. This species usually makes its appearance within the limits of the United States in October, and is most numerous, as already stated, on the Pacific side of the continent. It is often seen associating with other Geese, especially the Snow Geese, with which it appears to be on most friendly terms. The birds seek their feeding grounds, if away from the coast, in the early mornings, and as they often follow the same line of flight going and returning, many are shot by sportsmen who have taken positions along their routes. When much hunted they become very shy and wild, and permit nothing to approach them, and have sentinels posted to give due warning of danger, and as soon as an alarm is sounded each individual throughout the flock is on the alert, and if the cause of their suspicion remains, the entire company takes wing for another locality. Although the name by which this species is generally known to the gunners of the west is Brant, it has also various others in different parts of its dispersion. Some of these are Laughing Goose,—on account of its cry, supposed to resemble the sound man makes when laughing,—Prairie Brant, Speckled Belly, Speckled Brant, Gray Goose, Pied Brant, Yellow-legged Goose, etc.; and Oie Caille and Gray Brant in Louisiana. This Goose is a most excellent bird for the table, especially if young, as it receives from its customary diet no strong or disagreeable flavors, and can rank as an article of food with any other species of Goose, excepting possibly a young bird of the salt-water Brant. The downy young are very pretty little creatures, as they appear in their various colors of sooty brown relieved by olive and lemon yellow. This plumage lasts but a short time, when they begin to assume the mature dress, and early in the autumn they can hardly be distinguished from the adult, differing chiefly in not having the white on the head at the base of the bill, and less black on the lower parts.

In the Old World this Goose is dispersed throughout the northern portions, and ranges eastward as far as China and Japan. As is its custom in America it flies in V-shaped flocks, sometimes at a very considerable height, frequents low marshy districts and feeds upon water plants and grasses. At times it resorts to cultivated fields and picks up the grain scattered over the ground, but as a rule it is a vegetable feeder. It is not uncommon on the coasts of Great Britain, and in Egypt I found it the most abundant of the Geese that are accustomed to resort to the Nile. This species breeds near fresh-water ponds not far removed from the coast, depositing its eggs in a depression in the ground, lined with down. These are like those laid in America as may be supposed, yellowish-white, and six to eight in number. This Goose was well known to the ancient Egyptians, and its portrait frequently appears upon their monuments, and one of the earliest pictures of birds known to exist was found in a tomb at Mayoum, Egypt, and represents this species.

ANSER ALBIFRONS.

Geographical Distribution.—Northern portions of both Hemispheres, extending eastward to Japan. General throughout North America, south in winter to Cape St. Lucas, Mexico, and Cuba. Rare on the Atlantic coast. Greenland.

Adult.—Fore part of head, white, bounded posteriorly with a narrow, almost imperceptible, line of black. Rest of head and neck, dark brown; in some specimens the upper part of head and nape is very dark brown, causing this part to appear like a cap. Back and wings, grayish brown, feathers tipped with white. Greater wing coverts ash gray, tipped with white. Primaries, black. Rump, slate brown. Lower parts, grayish white, blotched with black, the amount of these blotches varying greatly among individuals. Upper and under tail coverts, white. Tail, dark grayish brown, the feathers edged and tipped with white. Iris, dark brown. Bill, orange yellow; nail, white. Legs and feet, orange or orange red. Average total length, 28 inches; wing, about $15\frac{1}{4}$; tarsus, $2\frac{1}{2}$; culmen, $1\frac{8}{10}$.

Young.—No white on the head, which is all dark brown, and no black marking on the under parts; nail of bill, dusky.

Downy Young.—Middle of crown and entire back, including

the upper surface of the wings and outer side of thighs, sooty brown, with an olive shade. From the bill a band extending back through the eye is of a slightly darker shade than surrounding feathers. Nape and back of neck, olive yellow. Entire lower surface rich lemon yellow, washed with lighter on the abdomen (Nelson).

BEAN GOOSE.

THIS is another species from the Old World, taken into our list of American birds on a statement that a specimen was seen or procured in North Greenland. However, this is not of much importance to those who shoot Wild Fowl, because it is not at all likely that they will ever meet this bird in the flesh in North America, and it is probably a very exceptional occurrence that one even goes as far west as Greenland. But there is a specimen in the zoölogical museum at Copenhagen stated to have come from that land, and on this testimony the Bean Goose becomes an American bird. We are not informed what are the reasons for believing the specimen came from Greenland, and museum examples have been known to bear wrong localities upon their labels, but let us hope this is not the case in this instance, and although we can never expect to see the Bean Goose flying free within our limits, it will be satisfactory to believe a venturesome individual did get at one time as far westward as Greenland. In many parts of Europe and Asia it is a common species, frequenting the coasts, and also inland localities more often than is usual with other species of Geese. It is a wary bird and keeps to open places, and has sentinels posted to warn the flock of approaching danger. It breeds in high latitudes.

ANSER FABALIS.

Geographical Distribution.—Northern Europe and Asia, in winter to southern Europe and Northern Africa. Very accidental in Greenland.

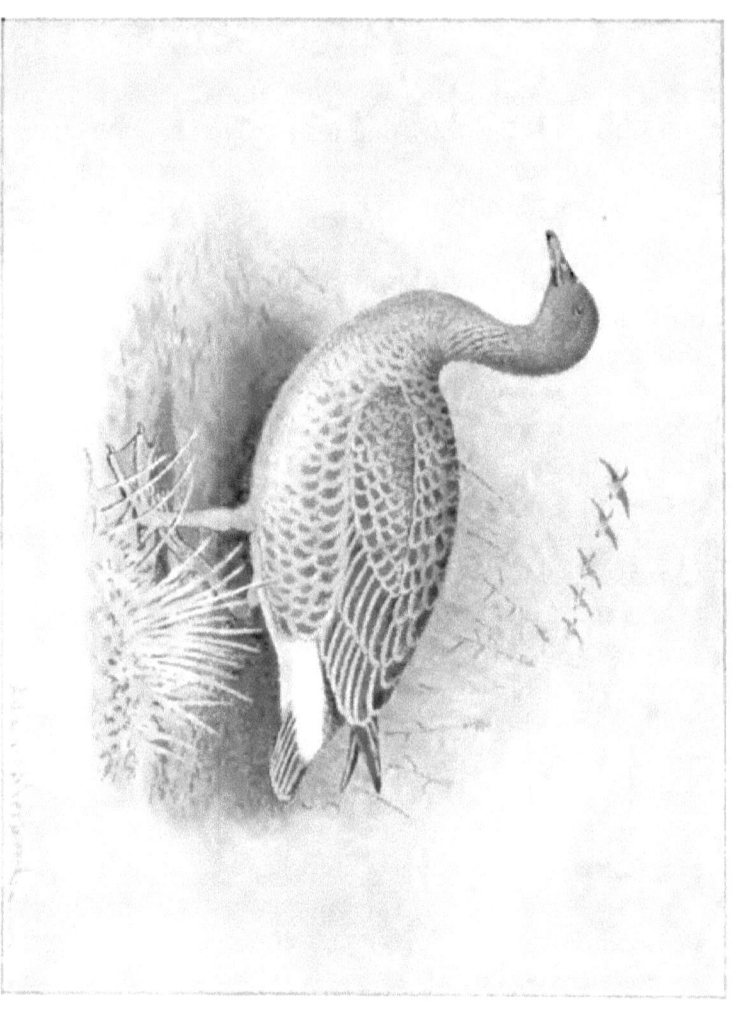

9. Bean Goose.

Adult.—Head and neck, grayish brown, darkest on head, white patch on forehead. Back and scapulars, dark brown feathers edged with grayish white. Rump, blackish brown. Wings brown, grayish on coverts, which with secondaries and tertials are edged with white. Breast, pale brown; sides and flanks, brown, edges paler. Upper and under tail coverts, abdomen, and vent white. Bill, middle part deep orange, remainder with nail black. Iris, dark brown. Legs and feet, orange. Total length, about 32 inches. Wing, 19; culmen, $2\frac{3}{10}$; tarsus, $2\frac{4}{10}$.

Female.—Like the male, but is usually somewhat smaller.

EMPEROR GOOSE.

THIS handsome Goose is one of the very few water fowl that are met with in North America that I have never seen alive, and on account of its very limited dispersion, one desiring to study its habits in its native haunts must visit that portion of Alaska lying between Behring Strait on the north and the Aleutian Islands on the south. This species breeds about the mouth of the Yukon, and around St. Michael's, and probably on the north coast of Siberia west of Behring Straits, and passes the winter about the eastern islands of the Aleutian chain. It is seldom seen within the limits of the United States, but occasionally a straggler is taken within our borders, as in the winter of 1884 when one was procured in Humboldt Bay, Northern California, by Mr. Charles Fiebig, who says the Emperor Geese occur there at long intervals.

Mr. E. W. Nelson, to whom we are indebted for much of our knowledge of the habits and economy of the various birds that periodically visit the Arctic regions, has given some interesting notes of this species, of which the following is a transcript. From the 22d of May to June 1 this Goose becomes daily more common at St. Michael's, until at the latter date the main body has arrived, and their forms and notes are as familiar as are those of the White-fronted and White-collared or Cackling Geese. The first comers are very shy, but become less so when they begin to arrive in flocks. At a long distance they can be distinguished by their heavy bodies,

D. G. Elliot.

10. Emperor Goose.

short necks, and quick wing-strokes, resembling those of the Black Brant. Although not so rapid on the wing as that species, nor in fact, as are other Geese, they are nevertheless swift flyers. When on their way between feeding grounds they utter a hoarse, deep, strident *Clâ-hâ, clâ-hâ, clâ-hâ*, very different from the note of any other Goose. Soon after their arrival mating begins, and in couples they fly about keeping close to the ground, rarely rising thirty yards above it. The males are jealous and pugnacious, making a vigorous onslaught upon any one of their kind or any other species of Goose, should they draw near. When a mated pair are feeding, the male is restless and watchful, and if alarmed the birds draw near each other, and before taking wing, both utter a deep ringing *U-lûgh, û-lûgh*. There is a peculiar deep hoarseness about this note impossible to describe. By June the females begin to lay on the flat marshy islands near the sea, and at low tide the broad mud flats on the shore are thronged with them, and after feeding, they congregate on the bars until forced to leave by the incoming tide. Most of the nests were placed on the marshes, and sometimes the eggs were deposited amid driftwood below high-water mark. It is not always easy to distinguish this Goose when on the nest, even when there is not much cover, as the bird extends her head and neck flat upon the ground, remaining perfectly motionless, and does not leave the nest until the object of her alarm has passed, when she usually moves off with a startled cry. The eggs are placed in a depression in the ground, and in number they range from five to eight, and when fresh are pure white or nearly so, but become a dirty brownish white after remaining in the nest a brief period. As the number of eggs increase, the female forms a bed of fine grass, leaves, and feathers, the latter plucked from her own breast. When disturbed the female usually flies straight away, sometimes for half a mile before alighting, and betrays little concern for her treasures. The male is rarely seen in the vicinity of the nest. By the last of June or beginning of July the young are hatched, and from the last of July to the middle of August the adults moult. At this season tens of thousands of Geese of all kinds are killed by the Eskimo, who set long nets across the marshes and drive the moulting birds into them. This slaughter is bad enough, but is rendered still more reprehensible from the fact that the savages kill thousands of young birds that are at such times entrapped, to prevent them, as they say, from being in the way for the next drive. Is it to be wondered that the Wild Fowl in North America are rapidly marching, in so many cases, toward extinction, when such practices are indulged in, even on their very breeding grounds? The eggs of the Emperor Goose are eagerly sought for both by the natives and whites, and take the place of meat on the daily bill of fare. When again able to fly, these Geese gather along the seacoast, and remain there until winter drives them to the Aleutian Islands a few hundred miles south. The natives south of the Yukon make dresses from the skins of this bird, as they do also of those of other species of Geese.

The Emperor Goose is difficult to kill, and it requires a heavy charge of shot to bring it down. It is hardly fit for food, the flesh being coarse, rank, and with a decidedly unpleasant odor, says Turner; but Dall states that though the flesh has an intolerable odor of garlic, which makes the process of skinning a very disagreeable task, yet this passes away when the bird is cooked, and he found it tender and good. This species visits the Prybilof Islands, but only as a straggler. In the month of October, usually from the 7th to the 20th, says Turner, a strong north-northeast wind blows, attaining at times great velocity. This has the effect of lowering the waters of Norton Sound to a remarkable degree, sometimes as much as eight feet below the lowest mark attained. At such times the Emperor Geese visit the vicinity of Stewart's and St. Michael's Islands to feed on the shell fish exposed by the receded water. By the 15th of November they depart for the south side of the peninsula and the Aleutian Islands, arriving at Unalaska by the 1st of December and remaining until the next April. The Russian name of this bird is Sa sar ka, which means Guinea Hen, as they fancy there is a resemblance in the coloring between that bird and this Goose. In the Aleutian Islands it is called the Lidenna Goose, and at Norton Sound it is known as the White-headed Goose.

PHILACTE CANAGICA.

Geographical Distribution.—Coast of Alaska, between Behring Sea and the Aleutian Islands. Mouth of the Yukon, possibly on Siberian coast, west of Behring Straits. Commander Islands, Kamchatka; casually in winter on the Pacific coast of the United States as far south as Humboldt Bay, California.

Adult.—Head and back of neck, white. Forehead and cheeks, frequently stained with rust color. Throat and fore part of neck, brownish black, feathers on lower part of neck, with a small white spot at tip. Back and under parts, bluish gray. the feathers having a subterminal black bar and white tips, much more distinct on the back than on the lower parts. Secondaries, brownish black, edged with white. Primaries, blackish brown. Lower back and upper tail coverts, bluish gray, the subterminal bar and whitish tips indistinct. Basal half of tail, slate color, remainder white. Iris, hazel. Bill, maxilla pale purplish, washed, with fleshy white; nail, horn white, edges dark horn color; mandible, horn color, with white spot on each side. Membrane of nostrils, livid blue. Legs and feet, bright orange yellow (Nelson). Total length, 26 inches; wing, $14\frac{1}{4}$-$15\frac{1}{2}$; tarsus, $2\frac{4}{10}$; culmen, $1\frac{4}{10}$.

Young.—Similar to the adult, but with the head and neck, brownish black; the feathers on top of the head, speckled with white.

11. Canada Goose.

CANADA GOOSE.

THE common Wild Goose is distributed generally throughout North America from the Arctic Sea to the Gulf of Mexico, and from the Atlantic to the Pacific Ocean, breeding as far south as Colorado, near lakes at high elevations. No species of our Wild Fowl is better known, nor its advent within our borders more eagerly anticipated. It breeds in many parts of the northern United States, and thence northward throughout the Arctic Regions, chiefly, however, to the east of the mountains. In Alaska it is rare upon the coast, though it is met with along the Yukon River, but is supplanted in that Territory by several allied though smaller species. It has been found nesting by Richardson on the lower Anderson River, but he says it does not go to the coast. It seems to prefer the interior of the country during the breeding season, selecting wooded and swampy districts, and apparently at that time avoids the neighborhood of the ocean. Its arrival in the northern latitudes from the South is always hailed with joy by the inhabitants of those cheerless regions, as they depend largely upon these birds for their means of subsistence. It is among the first of the Wild Fowl to appear in the spring, and soon begins to prepare for its matrimonial duties.

In about three weeks after their arrival the birds have selected their mates, and are dispersed throughout the country, choosing sites for the nests in secluded places in the vicinity of quiet water, and where the cover of grass or plants is sufficient for concealment. The nest is usually upon the ground, although it has been found upon the stump of a tree surrounded by water, and also in the branches of a tree at a considerable height. It is composed of various materials, such as dry plants, dead leaves and grass, or sticks and moss, lined with feathers and down, and is quite large. The eggs vary from six to nine, sometimes more, when the bird is domesticated, and they are a uniform ivory white. During July the young are hatched, and the old birds moult. This is a dangerous period for them, as their means of escape are limited to hiding away in the marshes, at which they are very skillful, or else keeping out in the center of lakes or other large bodies of water. Many, however, are killed at this period, and sometimes whole flocks are captured alive, of which fact Hearne relates an instance when some Indians drove into Fort Prince of Wales, on the Churchill River, forty-one old and young birds which were incapable of flying, and which were herded as easily as if they had been domesticated.

As the days begin to shorten, and ice to form upon the inland waters and along the borders of the sea, the Wild Geese commence to prepare for their journey South. Much conversation is indulged in, and doubtless the various routes are discussed, and instructions to the young given as to how they must behave in the trying times before them; as there is no doubt that birds and other animals can converse as intelligently with each other as men can, so far as making their wants and intentions known. Feathers having been thoroughly preened and cleansed, and protected by an abundant dressing of oil, everything is in readiness, and a favorable wind from the north having sprung up, the flock, usually consisting of a single family (although sometimes two or three may join together), with loud cries and much flapping of the wings, and beating of the water with the feet, rises in the air and takes a direct course for the winter home. Led by some experienced gander, who has also the extra duty of cleaving the way through the air, which becomes at times most fatiguing, the birds are strung out in a lengthened V-shaped line, each one protected to a certain extent against the wind, if adverse, by the one in front, and with slow, heavy beating of the wings, the flock speeds on by day and night with great rapidity.

> "Then stood we shivering in the night-air cold,
> And heard a sound as if a chariot rolled
> Groaning adown the heavens; and lo! o'erhead,
> Twice, thrice the wild geese cried; then on they sped,
> O'er field and wood and bay, toward Southern seas;
> So low they flew that on the forest trees
> Their strong wind splashed a spray of moonlight white;
> So straight they flew, so fast their steady flight,
> True as an arrow they sailed down the night;
> Like lights blown out they vanished from the sight."

There is nothing to intercept their course; in the great fields of air through which they move, there are no bounds or limits, nor barriers of any kind; the route is free and open. At least so it appears to us as we watch them steering across the blue vault of heaven, sending down at intervals from out the sky a note of recognition to the inhabitants of earth.

But all is not so free and without restraint, even to the voyagers of the trackless wastes of the airy regions, for in their path rises occasionally a fleecy mist that obscures all landmarks, and although it might be supposed that birds like these, whose instincts are so keen and unerring, would never lose the points of the compass, yet when shut in by a fog or encompassed by a storm of snow, the Geese become confused, seem to lose all knowledge of their course, and frequently descend and alight upon the ground. Passing over large cities, or forests of shipping, sometimes has a similar effect upon them.

Migration is performed usually at night, though at times many flocks are seen journeying by day. When desiring to rest and feed, the ground beneath is carefully scanned, in order to select the place offering the best sources of nourishment, as well as affording security from all danger. A suitable spot having been found, at a call from the leader the birds begin to descend, lowering themselves rapidly, and at times sailing along on motionless pinions. If they have decided suddenly to stop, they will frequently drop abruptly in a zigzag course, as is described in the articles on certain species of Ducks, and, when nearing the ground or water, turn against the wind and settle gently down.

When traveling the leader often utters a *Honk*, as if asking how those following him were getting on, and is answered with an " All well " reply from the rear. If he becomes fatigued by the extra labor of cleaving the air he falls out to one side, and some other old bird moves up and takes his place, the former leader dropping into the ranks again without disturbing their regularity or checking the speed. This movement is accomplished with an ease and smoothness that could only come from long practice, and is most pleasing to witness.

Toward October, or, if the season is late, some time in November, these Geese begin to arrive on the waters of our sea-coasts, and throughout the interior of the United States, seeking their winter quarters. They come in comparatively small flocks, succeeding each other rapidly, generally flying high in the air, and, on alighting, congregate together in masses, often containing many hundreds of individuals. They are usually very noisy, the *Honks*, in many keys and variations of inflection, resounding from every side. They seem delighted to have successfully reached what may possibly be the termination of their journey (though doubtless many a member of the little band has fallen by the way), and splash about in the well-known waters, wash and dress their feathers, and maintain an uninterrupted flow of conversation. They keep much to themselves, whether on the prairie or on the water, associating at times with the Swan, if any are in the vicinity; though they make no objection to flocks of various species of Ducks remaining with them, and it is no unusual sight, on large bodies of water in winter, to see flocks of Geese surrounded and mixed up with great multitudes of deep-water Ducks, and even Mud Hens or Blue Peters (*Fulica americana*), which on calm days are in the habit of gathering in large numbers on the open water away from shore.

At all times the Canada Goose is a vigilant and wary bird, having sentinels posted at various points when the members of a flock are feeding, which with outstretched necks remain motionless, keeping a keen watch around. These are not neglected by their fellows, but, after a spell of duty, are regularly relieved by others. While trusting in a large degree to their guardians, the other members of the flock are by no means neglectful of all proper precaution, and each one also is on the alert for danger even when engaged in feeding. They subsist upon berries in their season, grasses, roots, and leaves of various marine plants, which they dig up from the bottom with their bills. This Goose does not dive when feeding, but, keeping in shallow water, tilts up the hind parts as do the Mallard and other Ducks, holding itself in position by paddling with the feet, and reaching down to the full extent of the long neck, grasps and pulls up the tender grass and plants growing beneath. Sometimes the flocks dig large holes in the bottom, but commit nothing like the damage, nor waste such quantities of food as do the Swan. Canada Geese have no special time for feeding, and seem to find much pleasure in the occupation both during the day and night. If they desire to seek their food in the marshes, they generally enter them at night, two or three hours after sundown, and their arrival in such places is always known by the honking of the birds as they prepare to alight, or as those already on the ground salute the newcomers. While feeding, if feeling secure, they are often very noisy, and keep up a continual calling. Soon after the rising of the sun they leave the marshes and retire to the bays and sounds, and usually keep well away from the shore.

When a flock is on the wing, its members always give an intimation of their desire to alight by sailing on motionless pinions for a short distance. Unless frightened away, this action is almost universally the precursor of a cessation of flight. The Wild Goose is very fond of sanding, as it is called, and daily will visit the beach or bars in the rivers or sounds to obtain this much-desired article, and if undisturbed will gather in such places in immense numbers at certain stages of the water or tide. Advantage is taken of this habit by sportsmen, and holes are dug in the sand, into which boxes are placed large enough to hold one or two men, and sand piled about them as a breastwork, or surrounded by reeds stood upright. Wooden or live decoys are placed about this blind, according to the direction of the wind, for the Geese will always swing round so as to come up to them against the wind before alighting. A flock of these large birds approaching the decoys is a beautiful sight, and we will take our position in such a box and see how they appear as in all confidence they draw near the dangerous spot. The boxes are either long enough for a man to lie down in at full length, or deep and wide enough to enable him to sit upon a bench or plank nailed across it about halfway down.

We take our places in one of the latter kind, and look out through the reeds over the water. If we have live decoys they are strung out in diverging lines. each bird tied by the leg to his perch or post, on which is a platform just below the surface for him to stand on when tired with swimming. Before us stretches the wide expanse of the sound or bay, traversed at times by small skiffs, which, with their white sails, resembling birds' wings, dart hither and thither. Various kinds of Ducks are speeding along in undulating lines high in air. or just skimming the surface of the water, while with a whiz and a buzz, a Hooded Merganser, or Ruddy Duck, or Buffle Head will swing in toward our hiding place and then dart by at a speed an express train would be unable to equal. But moving slowly along apparently, on heavy wings, a dark mass comes into view, piercing the air with its wedge-shaped phalanx. At times a faint cry is borne to our ears, like a challenging note. and the decoys cease for a moment from struggling with their straps, or from preening their feathers, and with lifted heads stand motionless, listening for a repetition of the well-known sound. The flock, at first so indistinct, now is well in view, and the call of the leader, responded to by his followers, comes over the water in clear and unmistakable tones. The decoys are at once alert, and their ringing notes of invitation are uttered earnestly and in quick succession. The oncoming birds hear the call, and, catching sight of their brethren supposedly enjoying themselves in a most favorable location, turn in their course, and rapidly approach the spot with answering cries. As they draw near the decoys become silent, and the advancing birds also cease their calling, and even though members of their own race are standing in full view, with that wariness and suspicion which is their very nature, they gaze with watchful eyes about the place. Usually, seeing nothing but their own kind before them, and stillness reigning around, they set their wings preparatory to alighting.

Nothing in Wild-fowl shooting than this oncoming phalanx is more beautiful or attractive to the sportsman,—sitting like a stone image in his box, hardly daring to breathe, gripping his gun as if his fingers would sink into the metal of the barrels,—as he peers between his enveloping rushes. Onward they come, the birds floating on silent wings, at equal distances apart, looming up to the eyes of the stiffened gunner in his crouching posture until they seem as large as Swan, and twice as near as they really are. The decoys, as if they knew what would be the result of this arrival of their brethren, and (so like is bird nature to much of human nature), rather exultant at the success of their share in the deception, remain still and watch the approaching birds. Getting nearly abreast of the leading decoy, the flock swings around toward the wind and, facing the breeze, with a few flaps glide gently into the water. They now gather together in a bunch and, having satisfied themselves that they have nothing to fear, swim gradually up to the decoys, and frequently commence to fight with them, but finding that they are fastened to something, and some of the captives beginning to struggle for freedom, their easily aroused suspicions are awakened, and they begin to move away.

The sportsman, who has been waiting for a favorable opportunity to get as many heads in line as possible, so as to secure the most birds at the first shot, seeing this action, is obliged to accept the chance he can get before they swim out of gunshot, and aiming where the heads are thickest, without rising discharges his first barrel, and springs to his feet, to avail himself of the next best opportunity. With the roar of the gun, the Geese rise *en masse*, and the air is full of twisting birds and flapping wings, a mixture of varying strokes and moving forms most bewildering to the novice, who, distracted by the commotion, probably fires his remaining charge in the air, expecting most of the birds to fall. Not so the cool and experienced shot, who, knowing full well that he can only get a single bird, except by accident, selects the one giving the most favorable opportunity, and adds it to those floating on the water. The remaining Geese rapidly take themselves away from such a dangerous neighborhood, and with many *Honks* express their disapproval of the whole business. It is astonishing how speedily such large birds can get upon the wing and out of range on such an occasion as the one described. The decoys, which have remained quite silent during all the commotion, and have witnessed the slaughter of their brethren, now express their satisfaction by splashing the water over themselves, swimming about and gabbling to each other rapidly in low tones, and then mount onto their platforms to watch for more Geese to allure to destruction. The dead birds float back upward, if shot on the water, with the head and neck immersed, while the wounded ones, laying the head and neck flat upon the surface, try to skulk away, paddling toward the marsh or beach to hide, or directly in the wind's eye for the open water. It is wonderful how skillful wounded Geese are in getting away, and how difficult it is to see one skulking at any distance upon the water if it is at all rough. They can dive and go quite a little distance under the surface, and they avail themselves of all the artifices at their command, to escape capture. If a wounded bird succeeds in gaining the marsh or an extensive bed of reeds, nothing but a good retriever is able to capture it.

Sometimes when a flock has settled before the decoys and is swimming toward them, and the sportsman is getting ready to fire, a *Honk* is heard above, and another flock comes sailing in to join the others, thus necessitating a cessation of hostility for the time being. I remember on one occasion when, as I was about to fire at a number of Geese before me, I was stopped by hearing the call of an old gander as he led his company up to my blind, and he was succeeded by flock after flock arriving in succession in the same way, keeping me in a constrained, uncomfortable position, for I did not dare to move, the birds being both over and around me, until at least one hundred Geese were gathered in front of my position. It is such occasions that try the nerve of a sportsman, and compel him to exert himself and control his natural impulse to shoot at the many birds in close proximity, and patiently wait for the more favorable chance upon the water. The flight of the Wild Goose, though apparently labored, is really not so, and the bird moves at a rapid speed, and is able to protract it for a considerable length of time. The beat of the wings is steady and performed with great regularity, and their wide expanse is one of the causes of the fine appearance of the birds when sailing up to the decoys.

The Wild Goose is easily domesticated, and will breed in confinement, and often is as contented in captivity as the common farmyard bird. Individuals that have been wounded and captured, after they have recovered, often make excellent decoys for their wild brethren, honking with great vigor at every flock which comes in sight. They are easily kept in confinement, only evincing a desire to depart when the time for the annual spring migration comes, and then they watch for their brethren on the wing bound for the northern breeding grounds. In the interior the Wild Geese visit the grain-fields in great numbers, and many are killed in such places, from blinds made in the stacks of straw, or in holes in the ground. Also the latter device is employed out on the open prairie in the route the birds have adopted during the evenings and mornings, when flying to and from their feeding grounds.

As spring draws near and the green of the reviving grass and rushes, and the swelling of the buds upon the trees denote the beginning of another summer, the Wild Geese grow uneasy and congregate together, keeping up an incessant honking and gabbling, with much dressing of the feathers and general preparation for a great event. As the days lengthen and the sun grows warmer, a few flocks will be seen high in air, headed to the northward, and at length the time comes when, all being ready, the main body, with many *Honks* as in one great chorus of farewell, takes leave of its winter home, and starts on its long journey toward the Pole. Some linger on, keeping company perhaps with wounded birds unable to conquer the long route northward, and some remain to breed even in latitudes that may be considered southern. But after the month of April, in most localities, unless the season is exceptionally late, the great armies of this species have left our limits, and the sounds and bays and wide sheets of water, which during all the dreary months have echoed with the stirring calls, and been enlivened by the moving, active figures of these gamy birds, will lie silent and in many instances deserted, until with the chill winds of another autumn are heard the joyful cries of the returning squadrons, recognizing again their winter home.

This species has very many trivial names, and besides those already employed, is called by some Cravat Goose, Bay Goose, Black-headed Goose, Reef Goose, and Gray Goose, while in Louisiana it is known as Outarde.

BRANTA CANADENSIS.

Geographical Distribution.—Throughout North America, from the Arctic Sea to the Gulf of Mexico, and from the Atlantic to the Pacific Ocean. Breeds in Northern United States and throughout the Arctic regions, mainly east of the Rocky Mountains.

Adult.—Head and neck, black. A triangular white patch on each cheek, extending over the throat, sometimes divided on the latter by a black line. Upper parts, dark brown, the feathers tipped with light brown. Primaries, rump, and tail, black. Lower parts gray or brownish gray passing gradually into the white of the anal region. Upper and under tail coverts, white. Bill, legs, and feet, black. Iris, brown. Tail feathers from 18-20. Individuals vary greatly in size, but the average will be somewhat as follows: Total length, 38 inches; wing, 18; tarsus, 3; culmen, 2½.

Young.—Similar to adult, but the white cheek patches are speckled with black, and the black neck grades into the grayish hue of the upper part of the breast.

Downy Young.—Patch on occiput and upper parts, olive green; under parts, light greenish ochre.

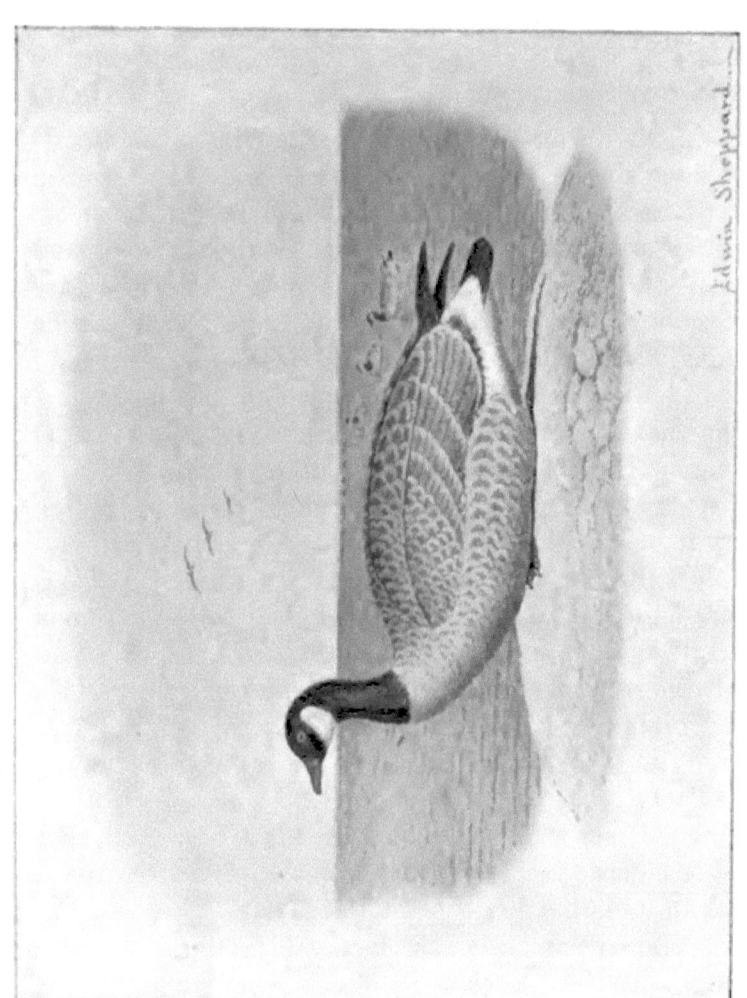

12 Hutchins' Goose.

HUTCHINS' GOOSE.

HUTCHINS' Goose during the winter season frequents chiefly the western portions of the United States. It breeds in the far north on the shores and islands of the Arctic Sea, and in the Delta of the Yukon, also at St. Michael's. It is abundant in the Aleutian Islands and nests on Atka and the Nearer Islands. The nests are placed on the shores near fresh water, or on small islands in the lakes or large ponds, and consist of a quantity of dry grass and leaves with some down and feathers intermingled. The number of eggs is generally six, and in the Aleutian Islands Dall says this species chooses hilltops for its breeding places, and the young were unfledged on July 10. In its habits and economy Hutchins' Goose resembles the Cackling Goose, but in appearance perhaps is nearest to the Canada Goose, though greatly inferior in size, its average total length being about ten inches less. In its migrations it usually keeps to the sea-coast, but in the United States it passes through the Mississippi Valley to the Gulf, but not in any great numbers, while on the Pacific coast it is one of the most abundant of the Geese. It associates with the Canada Goose, and once I shot a fine specimen of Hutchins' Goose from out a flock of its larger brethren at Puckaway Lake, Wisconsin. This specimen is now in the Museum of Natural History in New York. The flock was flying by, and noticing a small bird toward the rear of the line, I killed it, and found I had a fine specimen of Hutchins' Goose.

In California this species frequents the marshes on the coasts and also visits the plains in the interior, and joins the procession of Water Fowl as it moves, morning and evening, to and from its feeding grounds. The flocks are often approached by the sportsman, who keeps himself hidden behind an ox trained to walk slowly along, feeding as it goes, until their vicinity is reached and the gun can be discharged with deadly effect. Sometimes a wagon, drawn by oxen, can be driven near enough to bring the birds well within range. In Texas this Goose is also common, but upon the Atlantic coast is not frequently met with, so far as my experience goes. It may have been more common years ago in certain localities, and there may be others it occasionally visits at the present time, but I regard it as a scarce bird in the Eastern States.

Hutchins' Goose is known to sportsmen and baymen under various names, many of which are bestowed on account of its small size. Some of these are, Lesser Canada Goose, Small Gray Goose, Little Wild Goose, etc. It is also known as Bay Goose, Prairie Goose, Mud Goose, and Eskimo Goose in the far North; Winter Goose, Flight Goose, and Goose Brant. The specimens of this bird vary somewhat in their measurements, but the largest of them is only a miniature representation of the Canada Goose. The flesh of this species is excellent, and when the bird has become fat, feeding upon the tender grasses and water plants, it is a most desirable addition to a menu. The eggs are pure white in color, and of an oval form. Among the Aleutians this bird is called the Tundrina Goose.

BRANTA CANADENSIS HUTCHINSII.

Geographical Distribution.—Western North America from the Arctic Sea, through the United States from the Valley of the Mississippi to the Pacific, and south to the Gulf of Mexico. Rare on the Atlantic coast. Breeds on islands and along the shores of the Arctic Ocean and on the islands of the Aleutian chain.

Adult.—A small edition of the common Wild or Canada Goose, this bird is almost precisely similar in the color of its plumage, but is less in all its dimensions and has only from fourteen to sixteen tail feathers. The under parts are light brownish gray, gradually fading into the white of the anal region. The chin is black, but sometimes there is a white spot at the base of the mandible beneath. Like all the species of Geese the measurements vary considerably among individuals, but the largest Hutchins will rarely, if ever, equal in size the smallest Canada Goose. The number of tail feathers, however, will always serve to distinguish the two species. Total length will average about 30 inches; wing, about $16\frac{1}{4}$; tail, 5; tarsus, $2\frac{7}{8}$; and bill along culmen, $1\frac{1}{4}$. Tail feathers, 14–16.

WHITE-CHEEKED GOOSE.

THIS is purely a western bird, ranging from Sitka, in Alaska, along the Pacific coast to California in winter. It resembles very closely the Canada Goose, but the general plumage is perhaps a little browner than that of the commoner form; the white throat patches are separated in some examples, by a black stripe, and a white collar is around the lower part of the neck. This collar seems only to be possessed by birds in the fall and winter, gradually disappearing in spring, and becoming obsolete in summer. The habits of this subspecies do not differ from those of the Canada Goose, but its range is much more restricted. It does not appear to go north of Sitka, in Alaska, and was not seen around the Delta of the Yukon or vicinity of St. Michael's by any of the naturalists who have visited those districts. It is not improbable that this form is often found associating with flocks of the Canada Goose, and individuals may have been killed in various parts of our country, but as it would require an expert to distinguish them from the well-known species, and even if the white neck ring was noticed, it would probably be deemed an accidental occurrence and of no consequence, few instances of its appearance have been reported away from its usual line of migration. At St. Michael's Island this bird is called by the Russians the Lidenna Goose, the name given to the Emperor Goose on the Aleutian Islands.

13. White-Cheeked Goose.

BRANTA CANADENSIS OCCIDENTALIS.

Geographical Distribution.—From Sitka, Alaska, along the Pacific coast to California.

Adult.—Head and neck, black, the former having a large white patch covering sides of head and throat, sometimes separated by a black line on the throat, and extending upward to above and behind the eye. Chin, black. At the base of the black neck is a more or less distinct white collar. Back and wings, brown, lighter than in *B. canadensis*, with a grayish tinge, each feather tipped with white or brownish white. Primaries, black. Rump, black. Underparts, dark brownish gray, ending abruptly at the anal region, which, together with the upper and under tail coverts, is white. Tail, black. Bill and feet, black. Tail feathers, 18-20. Total length, 33-36 inches; wing, $16\frac{1}{4}$-18; tail, about 6; tarsus, $2\frac{8}{10}$; culmen, $1\frac{6}{10}$.

CACKLING GOOSE.

AMONG the Geese that frequent the Territory of Alaska during summer this species is the most abundant, breeding in great numbers from Point Barrow on the Arctic Ocean all along the coast to the mouth of the Yukon, and up the rivers into the interior; and also in the Aleutian Islands as far to the eastward, according to Turner, as Unalaska Island, beyond which it does not go. In winter it comes south to California, where it is abundant, and sometimes reaches the Mississippi Valley, having been taken as far to the eastward as Wisconsin.

It commences to appear in its northern breeding grounds toward the latter part of April, and the birds have usually all arrived by the middle of May. It is a great event not only for the Geese themselves, but also for the natives of the region, who have been living for many weary months on a diet of fish, and who welcome the opportunity to vary their monotonous bill of fare with the more generous article of flesh. Many birds are mated, Nelson says, when they arrive, but the males who have not yet succeeded in obtaining wives fight hard for the possession of the females. Nelson's description of these encounters is somewhat as follows. The females, keeping by themselves on the muddy banks of the river, a favorite resort, doze away the hours, or dabble in the mud. The males scatter about and are very uneasy, moving incessantly from place to place, and uttering loud cries. Occasionally two of these belligerently inclined

14. Cackling Goose.

birds will cross each other's path, when, uttering notes resembling low growling or grunting, each seizes the other's bill, and with wings hanging loosely by their sides, haul and twist one another, until suddenly coming close together, each strives to beat his rival with the wings, striking with so much force that the sound of the blows can be heard a long distance away. Not much damage is done, however, in these encounters, for the strokes are usually warded by the wing of the other bird, and the conflict terminates by the weaker breaking away from his antagonist and running off.

Mating having been at length accomplished, a spot for the nest is selected, generally a depression in a bunch of grass, or on a knoll, and this is lined with grasses or feathers plucked gradually from the female's breast, until the eggs are hidden in a bed of down. The number of these varies from seven to thirteen, and they are at first pure white, but after lying in the nest a while, become soiled and dingy. If anyone approaches the female when on the nest, she crouches down in as flat a position as possible, and when she deems it no longer prudent to remain skulks away through the grass, making no sound until she considers herself at a safe distance. In the latter half of June and the beginning of July the young appear, and are cared for by both parents until able to fly, which is toward the end of August. At this time the old birds moult. They now scatter over the country, feeding upon the different kinds of berries which are ripened throughout the land. On the Aleutian Islands, these Geese breed by thousands in the marshes and lagoons. On some of the Islands various species of foxes abound, and the Geese are compelled to rear their young on the islets near by, or on others in lakes, where they cannot be molested by their keen-witted foes. The female Cackling Goose is a persistent sitter, and will give up her life rather than desert her nest. Turner relates a circumstance which demonstrates this in the strongest manner. In the Islands of Agattu and Semichè, in the Aleutian chain, during the period of incubation, there occurred, in the latter part of June, a heavy snowstorm that covered the ground to the depth of three feet. The geese would not quit their nests and were suffocated, and the natives found scores of birds after the snow had melted, dead at the post of honor. The natives of Alaska capture many of the goslings of this species, and rear them, when they become very tame. When the weather is very severe in winter they require to be fed, but they also find a supply of food in a rather curious way. The roofs of the houses are covered with sod, and the heat of the dwellings causes the tender grass constantly to spring up, and the Geese are always on the housetops searching for these sprouts. The call of this Goose is a low *Honk*, or a rapidly repeated note like *Lŭck, lŭck*. A great number of these birds are killed during their stay in the North by all manner of devices, and are salted for winter use, the state of freshness of the meat at the time of packing being a matter of no consequence whatever, so long as it is Goose. Many are shot, others are caught in nets, and not a few are brought down by three or more stones fastened to thongs having their opposite ends tied together, and which revolve on being hurled into the air, and tangle up one or more birds in a flock flying low overhead.

This Goose begins to leave on its southern migration in October or beginning of November, according to the season or locality it is in. They are good judges of the weather and usually start before a storm. At times these birds arrive in California in October and remain until the following April. This species is the smallest of all the Geese, save Ross's, which enter the United States.

BRANTA CANADENSIS MINIMA.

Geographical Distribution.—Alaska; south in winter to California, and eastward occasionally in the Mississippi Valley to Wisconsin. Breeds in Alaska, and the Aleutian Islands as far west as Unalaska.

Adult.—This species is a small representative of *B. c. occidentalis*, and bears the same relationship to it as *B. c. hutchinsii* does to *B. canadensis*. The white patch on the head is rather differently shaped, and does not seem to go so far above the eye as in *B. c. occidentalis;* but this may vary in individuals, as undoubtedly does the amount of black on the throat. The main distinctions from the White-cheeked Goose, however, are size and the number of tail feathers, which in this species amount to from fourteen to sixteen, the same as in *B. c. hutchinsii*, but there are other and sufficient differences between the last species and *B. c. minima* which easily distinguish them from each other: such as the distinctive shade of coloration on the under parts, and its abrupt or gradual meeting with the white anal region. Sometimes examples of this species are strongly suffused beneath with rust color. Bill, legs, and feet, black. Total length, about 24 inches; wing, average, $13\frac{2}{3}$; tail, about $5\frac{8}{10}$; tarsus, $2\frac{1}{2}$; culmen, about $1\frac{1}{10}$; tail feathers, 14–16.

BARNACLE GOOSE.

THIS handsome Goose is a native of the northern portions of the Old World, and can only be regarded as a straggler into North America, and it is a doubtful question whether most of the examples that have been killed within our limits had not escaped from confinement, rather than were *bona fide* immigrants to our shores. The first one procured was at Rupert House, on the southern end of Hudson Bay, and was obtained by Mr. B. R. Ross. This was undoubtedly a straggler from Greenland, the southern end of which this species regularly visits. It has also been obtained in Nova Scotia and in Currituck Sound, North Carolina, that former paradise for Water Fowl. Long Island, also at one time a famous resort for all kinds of game, has yielded up one specimen. It would be difficult to name any species of bird that had ever visited the Atlantic seaboard an example of which had not at some time been procured on Long Island. This Goose is very abundant in various parts of the Old World, and resembles in its habits those of the Brant Geese of our own land. It feeds on grasses and plants, and can be readily domesticated and becomes as tame as the ordinary farmyard Goose. It is supposed to breed in Siberia, on the Tundras or barren grounds, and on the shores of the White Sea. It visits the Färöe Islands, Iceland, and Spitzbergen, and in its migrations is also found in the British Islands and many parts of the Continent.

It is a handsomer bird than the other allied species of Geese, and is about the size of the Brant. This species

15. Barnacle Goose.

passes much of its time on land, feeding on grass and roots, and it keeps up a constant gabbling both when occupied in feeding and also when on the wing, and is altogether a noisy bird. The eggs are said to be a uniform yellowish cream color. As is the case with our own Brant Geese, little is known about this bird's breeding habits or the localities it frequents at that season. It is called, sometimes, Bar Goose.

BRANTA LEUCOPSIS.

Geographical Distribution.—Northern parts of Eastern Hemisphere. Accidental in eastern North America.

Adult.—Head, nearly white; the lores, occiput, neck, and breast, black. Wings and back, bluish gray, feathers, with subterminal black bar, followed by one of white. Feathers of flanks, brownish gray, with white tips. Under parts, grayish white. Bill and feet, black. Iris, dark brown. Average total length, 25 inches; wing, 15; tarsus, $2\frac{3}{4}$; culmen, $1\frac{1}{4}$.

Young.—Cheek patch spotted with black. Feathers of back tipped with rufous, and wing coverts tinged with the same. Flanks barred with gray.

BRANT GOOSE.

THIS well-known bird is a native of the northern portions of both hemispheres, but in North America is found chiefly upon the eastern coast, and is rare in the interior, although at times it is met with in the Mississippi Valley. It is a bird of the salt water, and keeps to the sea, either on it, or near the inner side of the beach on the sounds and bays having an outlet to the ocean. It is not found on the Pacific coast, where it is replaced by the Black Brant, the succeeding species. The Brant breeds probably nearer the Pole than almost any other bird, its nest having been found in the most northern land yet visited by man. Captain Fielden found the first nest and eggs in latitude 82° 33′ N., and afterward many more in the vicinity. This Goose passes Hudson Bay in the spring and autumn in immense numbers, but makes no stop and is not seen in the interior, keeping always near the coast. The nests, which are mainly composed of down or feathers, are placed upon the beach near the water, but in Greenland, in Bellot's Straits, they are built in the cliffs which line the sides of this passage, according to the testimony of Dr. Walker, who saw this species in that place. The eggs are grayish white. During incubation the Gander remains in the vicinity of the nest, and when the young are hatched the parents conduct them to the lakes or open water near shore. The adults moult by the end of July. Brant make their appearance on the Atlantic coast of the United States in

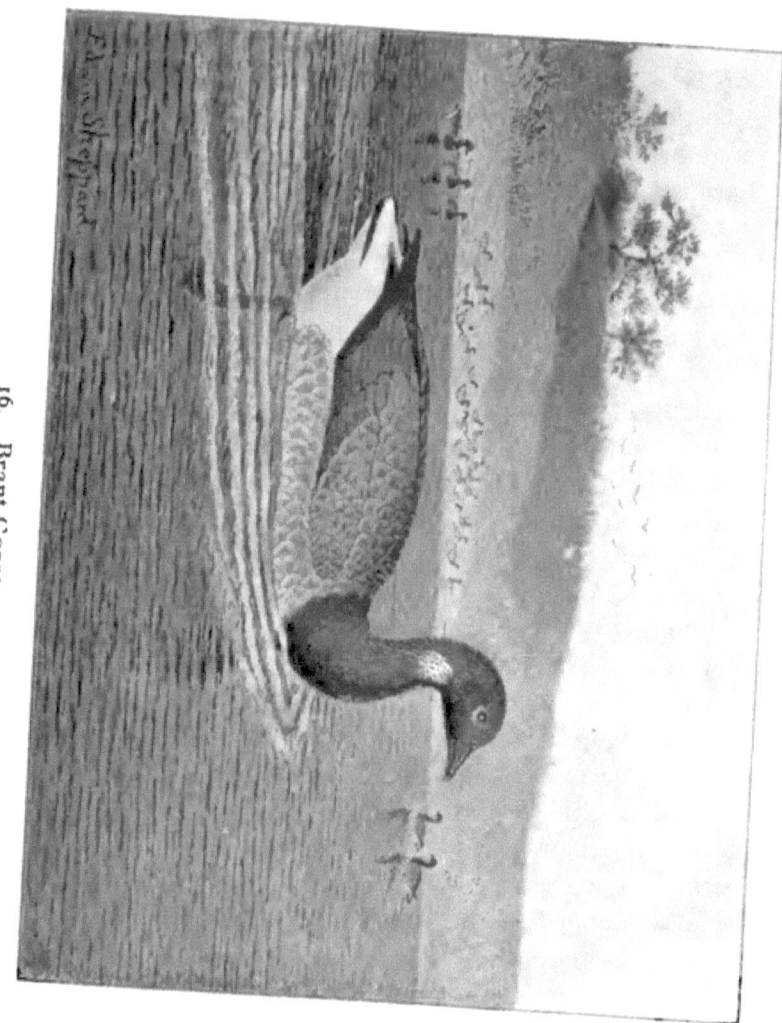

16. Brant Goose.

October, arriving in large flocks and congregating in chosen localities, sometimes in immense numbers. They fly in compact masses in a desultory sort of way, not very rapidly nor under any especial leader, and when in winter quarters rarely move far in any direction, and often return to the place from which they started. It has a peculiar guttural note, which is frequently uttered, resembling *car-r-r-rŭp*, or *r-r-r-roŭk*, or *r-r-rŭp*, and with a rolling intonation, and, when a large number of these birds are gathered together, the noise they make is incessant and deafening. I have been in the vicinity of a bar on which were congregated many thousands of Brant, and their voices made such a din that it was difficult to hear one's own in speaking, and when they rose at the report of a gun, the sound of their myriad wings was as the roar of rushing waters.

This Goose is usually very gentle, and when not much hunted pays little attention to man's presence. The birds come readily to decoys and are easily turned from their course by imitating their note, or by raising one leg or a hat in the air. As soon as their attention is attracted they swing around, and come to the decoys on motionless wings, in irregular, broken lines, uttering their rolling note, and if permitted, will settle down among their wooden counterfeits and commence to feed. I have known them try to alight upon the wings of my battery when I was in it, and the attending boat close by with sail up. They are easily killed, not nearly so tenacious of life as many Water Fowl, and, as they cannot dive, a wounded Brant is readily retrieved. It will skulk like other Geese with head and neck flat upon the water, and paddle away with all its might dead to windward, but it does not go very fast, and, if seen, is soon overtaken.

Brant are fond of sand, and it seems to be a necessity to them. Every few days the birds will resort to the bars in the sounds, or to the beach, and are often seen in such places standing in long lines or dense masses, dressing their feathers, or else sitting on the sand. When flying they keep over the open water, avoiding the land whenever possible, so that there is not often an opportunity given to shoot them from outlying points, or from a narrow strip between two bodies of water. As I have said, Brant do not dive, but feed in the manner of other Wild Geese, by tilting up the hinder part of the body and pulling up the grasses and roots from the bottom. Its food is the eel grass mainly, and although at times its flesh has a rather strong flavor, yet as a rule, especially in spring, it is an excellent bird for the table, and a young one is considered a delicacy. In calm weather Brant do not move about much, but gather in companies on the open water and feed, preen their feathers, or sleep, but before or after a storm they are uneasy, and generally in motion, flying apparently without any very definite purpose. But if the coming storm is likely to be severe, then they are seen flying, flock after flock, to some chosen place where they will be sheltered from the blast. In those situations at such times, the water is often black with the birds seeking a refuge.

Brant do not seem to be as plentiful in our eastern waters as formerly, constant warfare against them having greatly depleted their numbers, and in many places where they were once numerous they are now seen only in small bodies, or are absent altogether. This species has not many names, being almost universally known by the one at the head of this article, but sometimes it is called Brent, or Brent or Brant Goose, and also incorrectly Black Brant, which, however, is quite a different bird. In Spitzbergen, where it breeds, it goes by the name of " Ring-gaas," *i. e.*, " Ring Goose."

BRANTA BERNICLA.

Geographical Distribution.—Northern portion of both hemispheres. In North America mainly on the Atlantic coast; rare in the valley of the Mississippi.

Adult.—Head, neck, breast, and back at base of neck, black; a patch of white, in streaks, on either side of the neck. Upper parts, brownish gray, the feathers tipped narrowly with pale brown or grayish white. Under parts, grayish white, graduating into pure white about and under the tail. Middle of rump, brownish black. Upper and under tail coverts, pure white. Tail, black. Primaries and secondaries, brownish black. Bill, legs, and feet, black. Iris, brown. Total length, 24-30 inches; wing, 13; tarsus, $2\frac{4}{10}$; culmen, to end of nail, $1\frac{1}{2}$.

Young.—Similar to the adult, but with conspicuous white bars across the wings, formed by the tips of the coverts and secondaries. The white patch on neck is absent, and the under parts are uniformly lighter.

BLACK BRANT.

THE Black Brant represents the common species of the Atlantic shores on the Pacific coast, where it is no less abundant, and ranges from Alaska to California. It breeds on the shores of Alaska lying along the Arctic Ocean, about the mouth of the Anderson River, and westward possibly to the vicinity of Point Barrow. Numbers go still farther north, but where no man can tell; possibly to some unknown land amid the dreary expanse of the frozen Polar Ocean, which no human being has ever yet seen. These birds have been noticed in the autumn coming over the ice from the north to Point Barrow, which would make it fair to suppose that there was some unknown spot beyond the frozen barrier that was favorable for nesting and rearing the young, and other flocks of this Goose have been seen flying from the north to the eastward of Wrangel Land, and steering for the Alaskan coast, several hundred miles to the south.

The Black Brant is among the last of the migrants to arrive in Alaska in the spring from the south. It reaches St. Michael's and the mouth of the Yukon toward the latter part of May, and it takes about ten days or two weeks for the army of birds to pass, for none remain to breed; the goal they are steering for lying still far to the northward. It flies rapidly with quick, short strokes of the wings, not unlike those made by its eastern relative, and the flock, no matter what may be its size, is strung out in a single line at right angles to its course. Constantly waving, undulating movements run along the entire length of the line; commencing at either end, or

17. Black Brant.

from the middle, and going in opposite directions; caused by individual birds changing the level of their flight, and at a distance giving the impression of a shiver passing through the mass. This frequent graceful movement is very attractive to watch, and one keeps his eyes fixed on the birds, wondering where the next wave is going to begin. The same action occurs in the flight of certain species of Ducks. As a rule the Black Brant flies low. I think this is characteristic of the two species, and while the birds often change their altitude as they speed along, now just over the water, and again at no very great distance above it, they never rise to any great height. When flying, they keep to the coast line, following it in all its sinuosity, rarely passing over any part of the land, or else performing their migrations far out to sea. In the spring they are most abundant along the western Alaskan coast, but the birds are scarce in the autumn and must pass on their southward journey over the ocean out of sight of land.

Mr. MacFarlane, who found the nest and eggs of this species in Liverpool and Franklin bays, near the mouth of the Anderson River, and at various points along the shores of the Arctic Sea, says it was merely a depression in the ground, lined with a quantity of down. The number of eggs, which were a dull ivory, or grayish white color, was from five to seven, six being the usual complement. Some of these nests were placed on small islands in fresh-water ponds, and others on the shore or on islands in the two bays above mentioned. Some few individuals are said to breed on the shores of Norton Sound, in the marshes with Hutchins' Goose, but the great bulk of the migratory hosts pass on farther north. The Black Brant is a rare straggler to the Atlantic coast, and only a few individuals have ever been killed there.

It has occasionally been observed in the Mississippi Valley, and there is, I believe, a record of a specimen having been taken in Texas, but its occurrence east of the Rocky Mountains is extremely rare. It is common, however, along the west coast from Alaska southward in winter and generally keeps in the bays, or on the ocean a little distance from shore. It does not associate with other Geese to any extent, and does not go inland. The flesh of this Goose is tender and good, very similar to that of the eastern species, which it somewhat resembles in appearance. It feeds on marine grasses, and at times on small fish and crustacea. Beside the name of Black Brant by which it is commonly known, this bird is called on the Yukon, as stated by Kennicott, the Eskimo Goose.

BRANTA NIGRICANS.

Geographical Distribution.—Western North America, from the Arctic Ocean, at the mouth of the Anderson River, along the Alaskan coast. South in winter to Lower California.

Adult.—Head, neck, and upper part of breast, deep black; a broad white collar interrupted behind, on the middle of neck. Upper parts and wings, dark brown, nearly black on secondaries, primaries, and rump. Breast and abdomen, blackish plumbeous, almost as dark as the upper part of breast. Crissum, sides of rump, upper and under tail coverts, pure white. Tail, black. Bill and feet, black. Total length, about 25 inches; wing, $12\frac{1}{2}$; culmen, $1\frac{8}{10}$; tarsus, $2\frac{2}{10}$.

18. Wood Duck.

WOOD DUCK.

OF all the members of the Duck tribe scattered throughout the world, the present species is easily the most beautiful. The Mandarin Duck of China *Æx galericulata*), has a more bizarre appearance and is provided with curiously shaped feathers of various hues, and has altogether a most singular and unusual dress; but, though it may truly be considered a handsome bird, it cannot compete with this beautiful species, robed in a costume of harmonious colors so chaste and attractive as to find its most fitting expression in the name the bird possesses—the *Bride* of the Anatidæ.

The Wood Duck, Wood Widgeon, Branchier and Squealer, or Acorn Duck, as it is called in Louisiana, ranges throughout North America from Hudson Bay to the Gulf of Mexico, and breeds pretty much throughout its dispersion. It is a fresh-water bird, frequenting the lakes and rivers, often, also, resorting to swamps. On the seacoast, such as that of North Carolina, where, in Currituck Sound, the brackish waters and inexhaustible feed constitute a very paradise for Wild Fowl, the Wood Duck lives in the marshes, breeding on the mainland near at hand. It is one of the earliest of the water birds to start on its southern migration from the northern part of its habitat, leaving before the Blue-winged Teal, and often does not wait for the weather to become frosty, so anxious does it seem to be to get away from even the suspicion of winter.

The Summer Duck, as it is sometimes very appropriately called, breeds in hollow trees, and I have met with no instance when a nest was placed upon the ground. It will occupy the nest of some other bird in a hollow trunk, or will adapt some new-found cavity to suit its needs. It is astonishing to see how small a hole this duck can enter, and sometimes it approaches the opening to its nest, that appears not large enough to admit half the diameter of its body, but will pass in without difficulty. Usually the tree selected for the nest is close to the water, often overhanging it, but occasionally it may be a number of yards away. No matter how near the trees may grow together, or how thick may be the interlacing branches, the Wood Duck threads its way amid them with an ease and swiftness equaled only by a Wild Pigeon, and its flight is executed almost with the silence of an owl's in similar situations. This Duck appears to become much attached to its breeding place, and will occupy the same nest for successive years if it is lucky enough to escape the manifold dangers to which it is subjected. The nest is composed of grass, plants, and similar dried material, and is lined with down and feathers, mostly taken from the female's breast. A dozen or more white eggs, which soon become soiled, are laid, and then the male deserts his mate, and hies away to the society of other idle fellows like himself. The young, when hatched, are carried down to the water, one at a time, by the mother, in her bill, provided the distance is considerable, otherwise the little creatures scramble to the mouth of the cavity, and fearlessly drop themselves down into an element which they have never seen, but which their inherited instinct tells them is to be their future home. Whenever the female leaves the nest during incubation she always covers the eggs with the down and feathers so as to completely hide them, and thus insures a continuance of the warmth of which they are deprived by her absence. The young, when following the female, either upon land or water, continually utter a soft, low *Pee-pee*, a sort of prolongation of a chick's cry, and the mother answers with an equally gentle *Pēē-pēē*, something of the character of a whistle. Sometimes two ducks will take a fancy to the same nest, and much altercation then goes on, not so vociferous though as when the claimants happen to be both of different genera and species. A Wood Duck and a Hooded Merganser, as related by Brewer, contended for a nest, and fought continually for several days, and when the nest was examined it was found to contain eighteen eggs, all fresh, two-thirds of which belonged to the Wood Duck. The birds had been so persistent in their struggles to eject each other that neither had been able to sit.

This species is easily domesticated and breeds in confinement, provided it is afforded suitable locations for building its nest. It has a very gentle disposition and soon becomes tame and accustomed to new surroundings. It alights readily upon the branch of a tree, and also walks without difficulty upon the larger ones, and I have seen it alight upon the topmost rail of a fence surrounding a cultivated field, upon which it perched as comfortably, and seemed as much at home, as if it had stopped to rest upon the bosom of the lake which was close at hand. The Wood Duck, when moving over open water or marshes, in fact anywhere except in the woods, generally flies in a direct line, seldom altering its course or seeming to vacillate in its mind about the proper route to take. It flies swiftly, and when in the air looks a good deal like the Widgeon. It comes readily to decoys, and, if permitted, will alight among them.

Nothing in bird life can be much more beautiful than a

full-plumaged male Wood Duck, proudly swimming along, his lengthened crest slightly elevated, and the sun glancing upon the brilliant plumage with the metallic hues of green, violet, and purple scintillating in its rays. It seems to me that this beautiful bird has become scarcer in the past few years, and fewer return to well-known haunts. The beauty of the male makes him a desirable specimen for collectors, and the flank feathers are eagerly sought by the makers of artificial flies, while its flesh is always acceptable to the gourmands. Altogether, with so many suitors of various kinds, each desiring the bird for his own especial purpose, the Wood Duck's chance for becoming extinct is a very good one.

ÆX SPONSA.

Geographical Distribution.—Hudson Bay to Gulf of Mexico, and across the Continent within the above limits; Cuba. Accidental in Europe.

Adult Male.—Head, with a full, lengthened crest, almost reaching the back, of green, purple, and violet metallic hues. A narrow white line starts at the angle of the maxilla, passes over the eye, and extends to the end of the crest, widening slightly as it goes. Another broader white line commences below and behind the eye, and is continued along the lower edge of the crest. Behind the eye, and extending for some distance above the lower white line, is a broad patch of metallic purple. Cheeks and sides of neck, violaceous black. Crest, silky in texture of various metallic greens and purples. Throat and front of neck, pure white, with two falcate branches; the upper across the back part of cheek, to behind and nearly reaching the eye; the lower across the neck, going upward and beneath the crest almost to the nape. Back, dark brown, glossed with greenish bronze, the lower back and rump darker in hue, and grading into black on the upper tail coverts. Lesser wing coverts, slate brown, with a greenish gloss. Scapulars and tertials, velvety black, with rich metallic blue, green, and purple reflections, and the longest tertial is tipped with a white bar.

Middle and greater wing coverts, steel blue, with black tips. Primaries, slate color, changing to steel blue at their exposed ends, and with the terminal portion of the outer web, silvery white. Lower portion of throat and breast, extending onto the upper back, purplish chestnut, dotted in front with inverted V-shaped white spots, growing larger as they reach the breast. On sides of breast, above the shoulder of the wing, a broad black bar, above which is another of white. Sides and flanks, fulvous buff, crossed by fine, undulating black lines, the feathers on the upper borders having at their ends two crescentic black bars, inclosing a white one, the subterminal black bar being edged also on its upper side narrowly with white. Lower breast and abdomen, pure white. On each side of the rump is a patch of metallic dark purple. Some lengthened black upper tail coverts, with deep fulvous centers, fall over behind this purple spot. Under tail coverts, dark brown, grading into black at tips. Tail, black, with metallic green reflections. Bill, deep purplish red, becoming scarlet behind the nostrils, with a lengthened, pointed, black spot on the culmen, and the nail black. An oblong spot of white, from nostril to the nail, and the basal outline, gamboge yellow. Legs and feet, chrome yellow; webs, dusky. Iris, orange red; eyelids, vermilion. Total length, about 18 inches; wing, $9\frac{8}{10}$; tail, $4\frac{7}{10}$; tarsus, $1\frac{1}{2}$; culmen, $1\frac{3}{10}$.

Adult Female.—Head, plumbeous gray. Front, and a line on side of bill at base, space about the eye, extending backward to a point, chin and throat, pure white. Top of head and crest, the latter much shorter and thinner than the male's, glossed with metallic green. Back, rump, and upper tail coverts, hair brown, glossed with bronze and purple. Wings, similar to those of the male, but the secondaries widely tipped with white, and the speculum, metallic bronzy green, separated from the white tips by black. Breast, reddish brown, spotted with buff or buffy white. Rest of under parts, white. Flanks, umber brown, spotted with white. Tail, hair brown, glossed with bronze green. Bill, dark lead color, space on culmen, and nail black. Legs and feet, yellowish brown. Eyelids, chrome yellow. Iris, sienna. Total length, about 17 inches; wing, $8\frac{8}{10}$; tarsus, $1\frac{3}{10}$, culmen, $1\frac{3}{10}$.

Downy Young.—Top of head and upper parts, dark brown, darkest on head and tail. Sides of head, lores, and stripe over eye, bright buff; blackish brown stripe from eye to occiput. Spots on shoulder of wing, and on each side of rump, dull white.

BLACK-BELLIED TREE DUCK.

THIS species and the succeeding one are distributed through the countries lying south of the borders of the United States, and only enter a few of the Southwestern States contiguous to Mexico. The Black-bellied Tree Duck is not rare in certain parts of Texas in summer, along the lower Rio Grande, where it arrives from its more southern home in April. It is known there as the Long-legged Duck, and in Louisiana as the Fiddler Duck. When it flies it has the habit of uttering a clear whistling note that indicates its presence, especially at night, when most of its migrating is accomplished.

This species deposits its eggs in the hollows of trees, often at a considerable height from the ground, and the eggs, from twelve to sixteen in number, ivory white tinged with green, are laid upon the bare wood. The males leave the females when incubation commences and gather by themselves on the river, frequenting the sandbars, where they often congregate in large numbers. When the young appear they are carried to the water by the mother, in her bill. In the various countries lying to the south of our borders this Duck visits the grain-fields at night, especially the corn-fields, and commits considerable damage. It also frequents the swamps, and feeds on the seeds of certain aquatic plants, of which it is very fond. It perches easily on trees or on cornstalks, and its long legs enable it to walk and run with great ease and rapidity. It passes the day

19. Black-Bellied Tree Duck.

in the lagoons or other secluded waters, surrounded with woods or water plants, or sitting on the branches of trees, feeding and moving about mainly at night. It can be easily domesticated if taken young and is very watchful and will utter its shrill whistle at any unusual sound, or at the approach of any person on the premises. In some parts of Northern South America it is known as *Oui-ki-ki*, from its peculiar whistle, which is supposed to resemble those syllables, but in Mexico *Pe-che-che-ne*, for the same reason. Evidently it has a separate whistle for each country, or the idea of sound possessed by the people must be very different. A single specimen was procured by Xantus at Fort Tejon, Southern California, and this is the sole evidence of its presence in that State. Its dispersion seems to be mainly in the countries bordering on the Gulf of Mexico, from Texas, through Central America, and so on through the northern parts of South America, extending its range eastward to the West Indian Islands. It is a very pretty, gentle species, and the flesh, which is white and tender, is most excellent, indeed considered quite a delicacy. This Duck is by no means shy, and when domesticated keeps with the barnyard fowl, both day and night. It is a handsome bird, although its long legs deprive it of all attempts at a graceful carriage.

DENDROCYGNA AUTUMNALIS.

Geographical Distribution.—Southwestern States nearest to Mexico, and southward through Mexico, Central America, and northern South America; east to the West Indies.

Adult Male.—Forehead, pale yellowish brown; top of head, cinnamon; nape and line down back of neck, black. Sides of head and upper part of neck, ash gray. Chin and throat, grayish white. Rest of neck, upper portion of breast, back, and scapulars, cinnamon brown. Middle of back, rump, and upper tail coverts, black. Lesser wing coverts, olive ochraceous;

middle coverts, ash; greater and primary coverts, grayish white. The wing, when closed, shows a lengthened white or grayish white line for nearly its entire length. Primaries, dark brown. Tail, brownish black. Lower parts and sides of breast, yellowish brown, the cinnamon of the upper portion grading into this color. Abdomen, flanks, and under wing coverts, black; anal region, white, spotted with black. Under tail coverts, white. Bill, coral red; orange at base of maxilla. Nail, bluish. Legs and feet, pinkish white. Iris, brown. Total length, about 19 inches; wing, $9\frac{1}{2}$; culmen, $1\frac{9}{10}$; tarsus, $2\frac{1}{10}$.

Adult Female.—Resembles the male.

Young.—Similar to the adult, but colors duller. Abdomen and flanks, grayish white, barred with dusky.

Downy Young.—Superciliary stripe, and one over cheeks, encircling the occiput, bright buff; and one from cheeks to nape, blackish brown. Upper parts, blackish brown, with patches of deep buff, one on each side of back, and one on either side of rump. Underneath pale buffy yellow; belly, whitish.

20. Fulvous Tree Duck.

FULVOUS TREE DUCK.

WITH a much greater general dispersion than the last species, this Duck extends its range considerably farther north within our limits, and has bred in the marshes near Sacramento, California, and has also been found in Nevada, Louisiana, and Texas. In the latter State it is called the Rufous Long-legged Duck, and in Louisiana the Yellow-bellied Fiddler Duck, and Long-legged Duck, and it is abundant at times near Galveston. It is a summer visitor, like its relative, and frequents similar places. At the mouth of the Rio Grande this species is not uncommon and, it has been stated, it is also abundant at the entrance of the Nueces River. The Fulvous Tree Duck also breeds in trees, though the natives at Mazatlan affirm that it nests amid the grass. The eggs are pure white, and the female lays from ten to fifteen. This species resorts to fresh-water ponds or lakes, feeding principally upon seeds of grasses, and like its relative visits the corn-fields at night to obtain the grain. It is not wild, and affords much sport to the hunter, and its flesh being as tender and delicate as that of the Black-bellied Tree Duck, it is highly esteemed as an article of food. When wounded it exhibits such agility, running and dodging with so much speed, that it is very difficult to capture, and in deep water it dives and skulks with no little skill, and generally effects its escape. The plumage is not so attractive as that of the previous species, and it is a much plainer bird.

DENDROCYGNA FULVA.

Geographical Distribution.—States of Nevada, California, Texas, and Louisiana. Mexico, southern Brazil, and Argentine Republic. Accidental in Missouri and North Carolina.

Adult Male.—Top of head, deep rufous, darkest on the nape; sides of head, yellowish brown. A ring of black feathers, with white centers on middle of neck. A black line from occiput down center of hind neck. Lower part of neck, dark yellowish brown. Back and scapulars, black, broadly tipped with cinnamon, making these parts appear as if barred. Lesser wing coverts, chestnut; rest of wing, black. Tail, black; the upper and under coverts, white. Throat, buffy white. Upper part of breast, yellowish brown. Entire under parts, cinnamon. Flanks, with center of feathers, pale ochraceous, bordered with dusky. Bill, bluish black. Legs and feet, slate blue. Iris, brown. Total length, about 20 inches; wing, $8\frac{1}{2}$; culmen, $1\frac{3}{4}$; tarsus, 2.

Adult Female.—With the plumage very like that of the male.

Young.—Similar to adult, but little or no chestnut color on wing coverts. Under parts, paler, and the upper tail coverts margined with brown.

Downy Young.—A brown band from the ears to the hind neck, and one down the back of the neck. Occiput traversed by a white band, and one also across the wing. Upper parts, grayish brown; under parts, white.

21. Ruddy Sheldrake.

RUDDY SHELDRAKE.

IF it was stretching a point to admit the Smew among North American birds, when two females, it was claimed, had been taken in the flesh within our boundaries, what is to be said of this species' application for membership in our avi-fauna, based as it is upon two statements, one, that Dr. Vanhöffen, a member of an expedition to West Greenland sent by the Geographical Society of Berlin, reported that he saw a *skin* of this species in a collection of birds at Augpalartok in the District of Uppernavik, that was collected in that vicinity in 1892; and the other that, in 1895, Wenge of Copenhagen reports another specimen from North Greenland? These are the solitary instances of this bird's occurrence anywhere within what may be termed the limits of North America, which have been recorded. Doubtless Old-World species that breed in very high latitudes sometimes on the return journey go slightly astray from their regular course, and touch, possibly for a few brief moments, on some parts of boreal North America, and many more species probably do this than we shall ever know, but it is only to record an historical fact that any notice of these waifs and strays is taken at all, and they can in no way be considered as American birds.

This Duck is not, strictly speaking, however, a native of northern climes, but ranges in Southern Europe and Asia, and only accidentally goes to the Scandinavian Peninsula and Iceland. So rare is it in the north that, in the warmer climate of Great Britain, it is only a scarce straggler, and Dresser considers most of the specimens taken there have escaped from confinement. Still some of the rare stragglers to Iceland may have wandered farther, once they were off the right track, and reached Greenland. This species prefers the society of Geese to that of Ducks, and frequents, during the day, open fields where it can see a long distance, for it is habitually shy, going at evening to the lakes and ponds. It nests in the hollows of trees, also in holes in the ground and in clefts of the cliffs. It visits India, and my friend the late Dr. Jerdon related a legend of this bird that is current there. It runs that for some indiscretion two lovers were transformed into Braminy Ducks (the name for this species there) and were condemned to pass the night apart from each other on opposite sides of the river, and that all night long each in its turn asks its mate if it shall come across, but the question is always met in the negative: "Chackwa, shall I come?" "No, Chakwi." "Chakwi, shall I come?" "No, Chakwa." It is also supposed in some parts of India that whoever kills one of these Ducks will be doomed to perpetual celibacy; hence by the natives they are seldom molested. The call note of this bird is loud and clear, more resembling that of a Goose than any sound a Duck utters.

CASARCA CASARCA.

Geographical Distribution.—Southern and Eastern Europe; North Africa to Shoa, Southern Asia, China, and Japan. Accidental in the Scandinavian Peninsula, Iceland, and Greenland.

Adult Male.—Head and neck, buff, grading into orange brown on the lower part of the neck, which is surrounded by a black ring. Back, breast, and under parts, foxy red. Rump, yellowish red, vermiculated with black. Wing coverts, white;

secondaries, glossed with green and purple on outer web, forming a speculum. Tertials, yellowish, foxy red on outer web, gray on inner. Primaries, tail, and tail coverts, black. Bill, legs, and feet, blackish. Iris, brown. Total length, about 24 inches; wing, 14¼; culmen, 1⅝; tarsus, 2¼.

Adult Female.—Resembles the male, but the plumage is generally lighter, and there is no collar at base of neck.

MALLARD.

ORIGINALLY the source from which the domesticated races of Ducks have descended, the Mallard is distributed over the entire northern portions of both hemispheres. In North America it is found from the Arctic regions to the Gulf of Mexico, and from the Atlantic to the Pacific Ocean. Throughout this vast extent of country it bears, as may be supposed, many names, of which some of the most common are, Green-head, Wild Duck, and Gray Duck or Gray Mallard, while the French call it Canard français or French Duck; and the Russians Sé le sen. In England it is sometimes known as Stock Duck, probably because it is the stock from which the tame Duck has been derived. Wherever found in summer, there the Wild Duck breeds. The nest is a rather large structure of grasses and small sedge stalks, lined occasionally with down or feathers, and placed in the vicinity of water, in a marsh, or, if in the West, on the prairie near some slough. The pale, greenish white eggs are usually six in number, and the female alone attends to the duties of incubation; the male loitering about in the vicinity, or else joining unto himself a number of other idle males, passing the time in dabbling about the ponds in the vicinity and selfishly caring only for their individual interests. The female is a close sitter, and will allow an intruder to approach very near before indicating by any movement that she is aware of his presence, and only leaves the nest when capture is imminent. In the North the situation of the nest is sometimes quite different, and

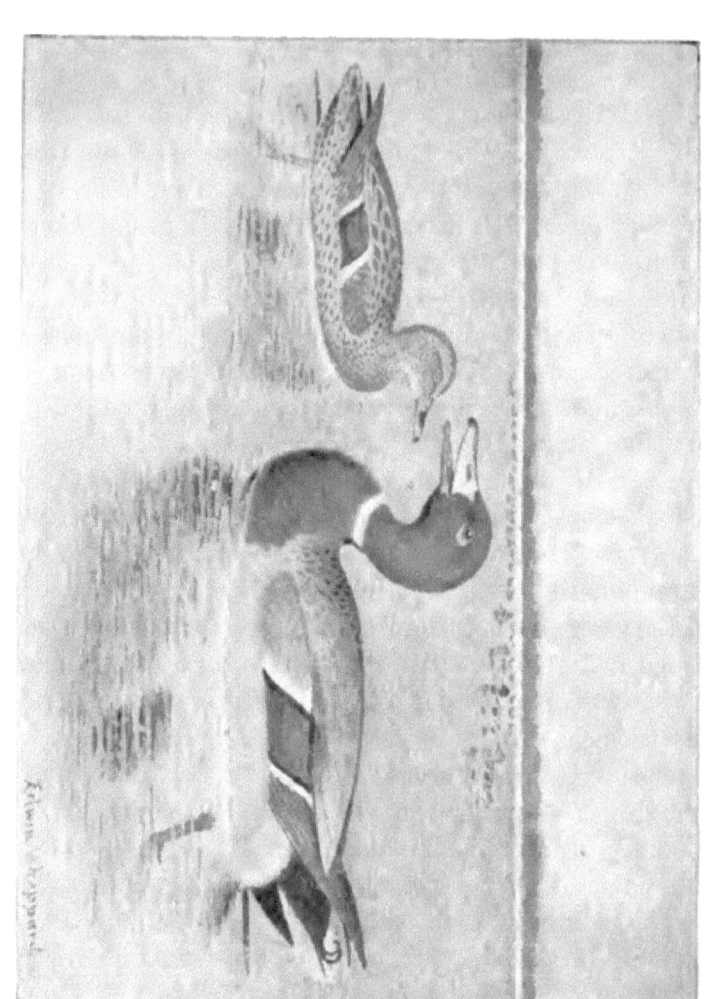

22. Mallard.

it is frequently placed among trees, occasionally in a hollow stump, even in the tree itself; usually, though, in such cases, the bird occupies some empty nest left over from a previous season. It requires about four weeks for the eggs to hatch, and the female at once leads the young to the water, and assists them to procure suitable food. The ducklings are very active, dive with ease, and hide at the least alarm with great celerity and success, sinking in the water and leaving the bill only above the surface. Numbers perish from various causes, for they have many enemies of the air, land, and water: hawks and owls, sometimes crows, also all kinds of four-footed creatures prowling about the swamps and marshes, not counting sundry snakes, prey upon them, while many a downy young disappears suddenly from the midst of the little family swimming quietly along, seized from beneath the surface by some turtle or predatory fish. So greatly are they exposed to manifold dangers that the only wonder is so many reach maturity. During the nesting season the males moult, the females not undergoing this process until the young are hatched. The breeding season is over by June, and when the brood is full grown the male rejoins his family. During the earlier part of the summer the plumage of the male is very similar to that of the female, but toward September he assumes the beautiful dress by which he is so well known throughout the world.

In the Northeastern States the Mallard is less common than farther south, and from New Brunswick to Massachusetts it may be considered as rather rare in comparison with other species of Ducks. In the Western States the Mallard visits the corn-fields, and in the Southern Atlantic States the rice-fields, and becomes very fat on these kinds of food, and also of excellent flavor. Advantage is taken of this habit by gunners, who make blinds in the fields where they can remain concealed and shoot the birds as they come in to alight or when flying overhead, and great numbers are killed in this way. Mallards also decoy easily, either to wooden counterfeits of themselves or to the bodies of their kindred that have been shot and set out before the blind, supported on sticks so as to give them a semblance of life. Usually wary and suspicious, it is often surprising to witness the entire confidence displayed by this Duck when approaching the decoys, particularly if the quacking notes in their various modulations are well imitated. On catching sight of their supposed relatives, the birds wheel, and come directly toward them, setting their wings as they draw near, and uttering low, soft quacks in a confidential tone, as if expressing satisfaction at meeting so many of the brethren at one time. Then, if any breeze is blowing, just before alighting they wheel head to wind and settle upon the water, but if it is calm they hover for a moment over the decoys and then drop with a splash in their midst.

When startled, the Mallard springs directly into the air several feet upward, and then flies away very rapidly. No preparation whatever is needed for it to make an exit from any spot, and if it is on a pond or narrow creek or in any concealed spot, one spring carries it above all obstacles and leaves a clear line of escape. Usually the sexes are not separated during the winter, but keep together, yet in North Carolina I have on several occasions discovered as many as fifty males assembled on a pond, without a single female being present. I have often wondered at this, and tried to account for such a concourse of one sex at that season of the year, about December, but never could arrive at any satisfactory explanation. The Mallard walks with ease, and can also run with considerable speed. On the water it moves with grace, and when seeking the seeds, roots, mollusks, various grasses, etc., on which it feeds, tilts up the hinder part of the body and digs on the bottom with its bill. It never dives, and when wounded tries to skulk away; perhaps as a last resort struggles to disappear beneath the surface, with, however, but poor results.

The Mallard is a very noisy Duck, and its loud quacking is one of the familiar sounds heard in the marshes during the winter. It is also very sociable and the little companies keep close together as they swim along, for even when feeding the birds rarely separate from each other for any distance. They are continually in motion, poking their bills into the soft mud, and sifting it through the mandibles. They feed mostly at night, but at the same time are equally active by day, although, if the weather is warm and calm, they are in the habit of taking a nap in the sun's rays, having one or more of their number, however, to act as sentinels and announce any approaching danger.

The Mallard is one of the commonest of our Water Fowl, and, from its large size and generally well-flavored flesh, is eagerly sought after. This Duck interbreeds with other species, and hybrids are frequently shot, bearing unmistakable evidence of their mixed parentage. Some of these are very beautiful birds, and in the days when hybridism was little understood or suspected, certain ones were described as distinct. One of these, and perhaps the most beautiful of all, was called by Audubon Brewer's Duck (*Anas breweri*) probably a cross between the Mallard and the Dusky Duck. Occasionally along the Atlantic coast a Duck is shot that is larger than the Mallard, with the head and part of the neck black with green reflections, and the lower portion of neck in front often white. Sometimes there is some white on the throat and head. Breast, very dark chestnut, under parts white, except the crissum, which is chestnut black. Back, brownish black variegated with grayish brown; rump and upper tail coverts, black with green reflection, like the head. These birds were regarded always with much interest, and opinions differed as to what they could be, but it is now generally considered that they are hybrids of the Mallard and Muscovy,* which, although bred in captivity, have returned to the wild state. The description given above only relates to one style or phase of the plumage exhibited by these birds, as individuals vary considerably from each other.

ANAS BOSCHAS.

Geographical Distribution.—Northern portions of both Hemispheres. In North America, ranging from the Arctic Regions to Panama and to Cuba. Breeding wherever it may be at the proper season.

Adult Male.—Head and neck, metallic green. White collar at base of neck. Back, brown, waved with narrow lines of pale brown. Scapulars, grayish white, waved with dusky. Wing, slate brown, edged with rufous on some feathers. Speculum, or wing patch, metallic purple, crossed at each end with a black bar, succeeded by a white one. Primaries, dark brown, with a grayish gloss. Lower back, and upper tail coverts, greenish black. Recurved feathers above tail, black. Breast, deep, glossy chestnut. Under parts, silvery gray, waved with narrow

---

* The Muscovy (*Cairina moschata*) is found throughout tropical America, and very possibly may visit at times the coasts of some of our Southern States, straggling outside its limits, and should it meet with the Mallard at the proper season, a mixed brood would very probably result. Therefore, some of these large ducks that are killed from time to time may not have been the offspring of domesticated parents.

lines of black, darkest on flanks and beneath the chestnut on breast. Under tail coverts, jet black. Tail, white. Bill, greenish yellow; nail, black. Legs and feet, orange red. Length, about 22 inches; wing, 11; tail, $4\frac{1}{2}$; tarsus, $1\frac{7}{10}$; culmen, $2\frac{3}{16}$.

Adult Male, when Moulting.—This stage of plumage occurs in the summer, and only lasts for a comparatively brief period, and is very like the dress of the female, but darker.

Adult Female.—Feathers of head and neck, with dusky centers and buff edges. Chin, whitish; throat, buff, or ochraceous. Upper parts, black; the feathers edged and tipped with buff on back and wings, and with ochraceous on lower back and upper tail coverts. Speculum of wing, as in the male. Under parts, buff, palest on breast and belly, with central streaks of black, broadest on sides and flanks. Bill, feet, and legs, colored like the male's. Dimensions, similar to the male's.

Downy Young.—Upper parts, olivaceous. Sides of head, stripe over the eye, and lower parts, yellowish buff, lightest on belly. A dusky streak from bill through eye to occiput, and a dusky spot on ear coverts. Pale buff spots on wing and on each side of back and rump.

DUSKY DUCK.

BLACK Duck, Black and Dusky Mallard, Black English Duck, and Canard Noir in Louisiana, are the names by which this bird is variously known. Its range is mainly throughout eastern North America, north of Florida, extending westward to Utah and Texas, and north to Hudson Bay. In Florida it is replaced by a smaller subspecies of similar appearance. In its habits this duck very closely resembles the Mallard, and it has the same loud quacking note. It breeds in various parts of the United States from Maine to Texas, as well as in Labrador, where in summer it is very abundant. The nest, placed upon the ground in the vicinity of water, is a compact structure of weeds and grass, lined with down and feathers, and the eggs are grayish white with a green tinge. Eight to ten is the usual complement.

Of all our Water Fowl the Black Duck is one of the most cunning and suspicious. It also possesses a keen smell, and no matter how well one may be concealed in a carefully constructed blind, if the wind blows toward the advancing bird, it will detect the sportsman's presence and remove itself without delay from the dangerous neighborhood. Many a time have I watched one or more of these wide-awake birds coming straight to my decoys, apparently only intent upon joining the flock of their supposed brethren, and uttering as they came that low, soft quack, so indicative of confidence and pleased satisfaction, when suddenly, without any apparent reason, the birds would rise in the air and swerve off in an

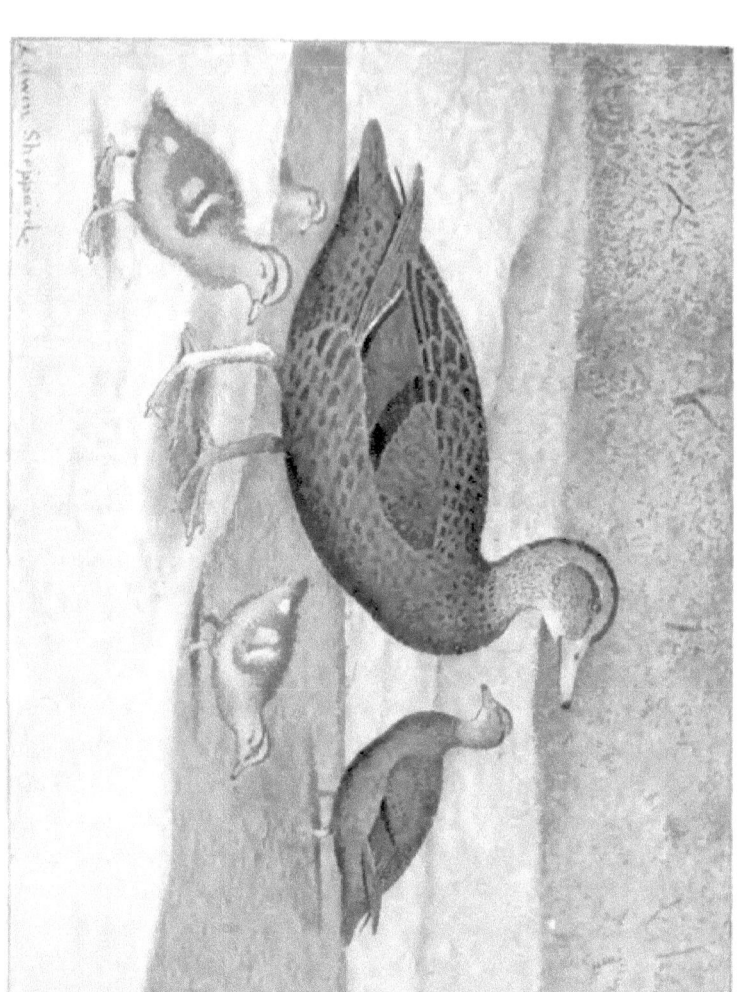

23. Dusky Duck.

opposite direction. There was nothing visible to create alarm, but their keen scent had warned them of the presence of an influence not accustomed to bring them increased happiness and a long life.

The flight of the Black Duck is performed in a similar manner to that of the Mallard, with quickly repeated beats of the wings, and usually at a considerable height, and as the bird moves speedily along it turns the head from side to side, sharply observing the ground beneath, and keenly attentive to every object and movement. When flying, the white under coverts of the wings show very conspicuously. It is usually on the alert, although at times its suspicions seem to be allayed for the moment, and then it will fly to the decoys and settle among them as quietly and with as much unconcern as would a tame Duck that was able to fly. These occurrences, however, are rare and not to be counted upon, as this Duck's trust in man is not often exhibited to any considerable extent. It rises from the water with a bound, as if it had been shot up by powerful springs, usually uttering a few quacks as it mounts upwards, scattering showers of spray around by the violence of its movements.

The Black Duck is very much of a nocturnal bird, moving about a great deal at night, especially if the moon is shining, and it associates with the Mallards and other swamp and marsh Ducks, its watchfulness and ability to detect danger making it a valuable member to any web-footed coterie. Its note is so like the Mallard's that it is difficult to distinguish them apart, and every few moments the quacks are shot forth in abrupt vociferations, as if the bird had just reached the limit of its power for suppressing them, and the voice had gained strength and sonorousness by long confinement. This species is a mud Duck, and delights to paddle and feed in the swamps and marshes, sifting the half-liquid ooze with its bill, and extracting whatever nutriment it contains, be it of plant, insect, or mollusk life. It is not particular as to its diet, and swallows anything it may find that is eatable. The flesh of this bird is not usually as palatable as is that of many other Ducks, although the quality varies of course with that of its food, but sometimes it is decidedly rank and fishy. It is a large bird, equal in size to the Mallard, and the sexes resemble each other very closely. Like the common Wild Duck, this species goes in flocks without any regular order, each bird selecting his own route totally regardless of his fellows, and frequently they present a confused mass in the air. Again, if over ponds, they wheel occasionally with some degree of unison.

ANAS OBSCURA.

Geographical Distribution.—Eastern North America, from Labrador to Florida; and west to the Valley of the Mississippi. Breeding throughout its range.

Adult Male.—Top of head and line on hind neck, black, streaked with buff. Rest of head and throat, buff, streaked with dusky. Remainder of plumage, dusky or brownish black; paler beneath, all the feathers, save those on lower back and rump, margined with ochraceous. Speculum, metallic violet, sometimes green, edged with black. Bill, yellowish green; nail, dusky. Legs and feet, orange red; webs, dusky. Length, about 22 inches; wing, 11; culmen, $2\frac{8}{10}$; tarsus, $1\frac{8}{10}$.

Adult Female.—Resembles the male. Practically there is no difference in the plumage of the sexes.

Downy Young.—Top of head, hind neck, and upper parts, olive brown; rest of head, neck, and lower part, darkish buff, lightest on belly. A dusky streak from bill through eye to occiput, and a dusky spot on ear coverts. Pale buff spots on border of wing, and on each side of back and rump.

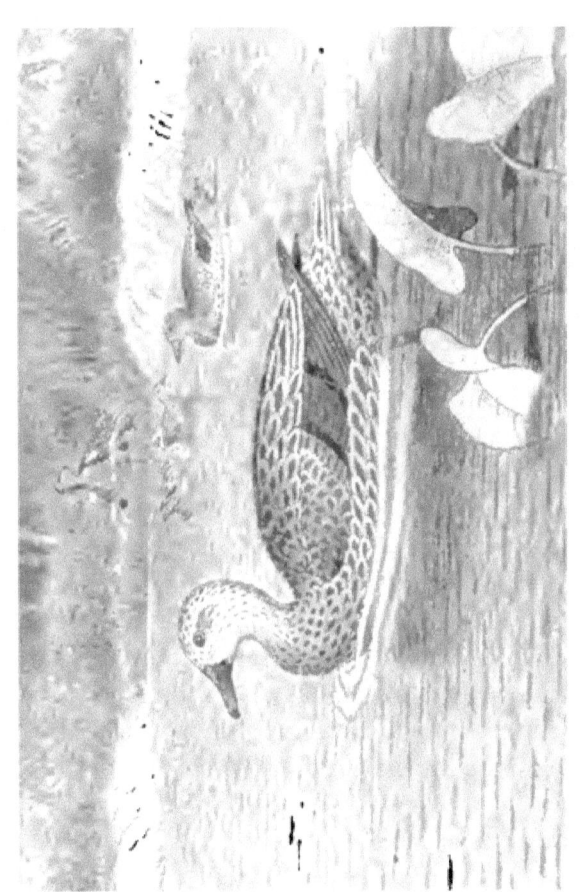

FLORIDA DUSKY DUCK.

THIS small representative of the Black Duck is apparently restricted to the more southern parts of the Peninsula of Florida. It is lighter in color and has a creamy buff throat and fore-neck. The bill is also differently marked and colored. It breeds in April, and the nest, formed of grass and similar materials and lined with down and feathers, is placed upon the ground in the midst of matted grass, or under a palmetto, or some sheltering bush, near water. The eggs, usually eight or ten, are very similar to those of the Black Duck, but lighter in color. The male remains in the vicinity while the female is incubating the eggs, but does not share in any of the duties.

This species frequents the ponds of fresh water, going out at night to the sheltered bays near the Keys to feed and disport itself. In the autumn the males appear to associate together, but flocks of both sexes are met with in the winter, and the mating season begins as early as January. Many are destroyed when the grass is burned to permit the young shoots to spring forth, as this is done usually at the period when the female is on her nest. In its habits this species does not differ from its Northern relative, is about as shy and cunning, but from its restricted dispersion and the number of sportsmen who visit Florida in winter, it has a very fair chance at no distant day of becoming extinct.

ANAS FULVIGULA.

Geographical Distribution.—State of Florida.

Adult Male.—Top of head, streaked with black and buff. Rest of head, sides, and back of neck, buff, streaked with dark brown. The cheeks are sometimes without streaks, but in a series of these birds plain cheeks were no more frequent than those with streaks, and this marking seems to be very variable. Chin and throat, plain buff of varying intensity. General plumage, black, feathers edged with ochraceous on upper parts, but with pale buff beneath. Speculum seems to vary in color among individuals, and is either metallic green or metallic blue, and, in some specimens, is tipped with white, forming a bar across the wing. Bill, yellowish olive; nail, and spot at base of maxilla, black. Legs and feet, pale orange red. Iris, brown. Total length, about 20 inches; wing, 10; culmen, $2\frac{1}{10}$; tarsus, $1\frac{8}{10}$; bill, 2.

Adult Female.—Resembles the male in general color of plumage, but is rather lighter, with sometimes a white bar across the wing on posterior edge of speculum. The legs and feet are dull red; the webs, flesh color, mottled with brown. There is little or no difference in the measurements of the sexes.

25. Mottled Duck.

MOTTLED DUCK.

THIS Duck was described by Mr. Sennett from a specimen taken at Nueces Bay, near Corpus Christi, Texas, by Mr. J. A. Singley, who was collecting birds for him at that time. It resembles closely the previous subspecies, the Florida Dusky Duck, but chiefly differs in having the cheeks streaked with brown, instead of being plain buff; and the speculum, or metallic spot on the wing, purple instead of green. The general effect of the coloration of the plumage is that of being spotted instead of streaked, and the light markings are pale buff instead of a deep buff, and this gives a slightly different appearance to the two forms, but they nevertheless resemble each other. The streaked cheeks are to be seen among some individuals of the Florida Dusky Duck, and the color of the speculum is at times merely a question of light, purple and green in metallic hues being often interchangeable. An ornithologist might readily recognize to which form most of his specimens belonged, but the ordinary observer would probably have difficulty in distinguishing them.

There appears to be a great similarity in the habits of this bird and those of its relative, as might be expected, but not many specimens have as yet been obtained, and more information regarding it is needed before the validity of its subspecific standing is satisfactorily determined. In Louisiana it is known as Canard Noir d'Eté, or Black Summer Duck. It is said to be a common resident in that State, and breeds there.

ANAS FULVIGULA MACULOSA.

Geographical Distribution.—Eastern Texas, Louisiana, north to Kansas.

Mr. Sennett's description of this bird is as follows:

"Top of head, blackish brown, margined with very pale buff. Chin and throat, isabella color. Cheeks, buffy white, with narrow streaks of dark brown. Feathers of breast, wings, upper parts, and flanks, blackish brown, margined with pale buff. Under parts, buffy white, each feather with a broad blackish brown mark near the tip, giving a decidedly mottled appearance. Under tail coverts, blackish, with outer margins of inner webs reddish buff; those of outer webs, buffy white. The four median feathers of tail, blackish brown; the others, fuscous, margined with pale buff, and a V-shaped mark, as in *A. fulvigula*, but of a buffy white. Under surface of all tail feathers, light gray, excepting the four median, which are blackish brown. Lining of wing, white. Speculum, metallic purple, feathers tipped with white. Bill has a small black spot on base of lower edge of upper mandible, as in *A. fulvigula*. Feet, reddish orange. Wing, 10 inches; culmen, $2\frac{1}{4}$; tarsus, $1\frac{5}{8}$; middle toe and claw, $1\frac{1}{2}$."

26. Gadwall.

GADWALL.

ESSENTIALLY a fresh-water bird, this Duck, while met with generally throughout North America, is nowhere so abundant as are the Widgeon, Sprigtails, Mallards, etc., with which it is accustomed to associate. It has a wide dispersion, and is found throughout both the northern hemispheres. In North America it is known by various names, those most commonly employed being, Creek Duck, Speckle-Belly, Gray Duck, Welch Drake, German Duck, Gray Widgeon, and Canard Gris in Louisiana. It is a shy bird, retiring in disposition, keeping to the small creeks, borders of marshes, and fresh-water ponds. It is a very swift flyer, and resembles very much the Widgeon when in the air, and dives with equal celerity and address. It hides among reeds and tall grasses and passes much of its time seeking its food close along the shores, where for the greater part of the time it is concealed by overhanging bushes or grasses. Generally it goes in small flocks, does not readily come to decoys, and when it does draw near them it is probably in the company of a small flock of Widgeon. The Gadwall breeds in the United States, as far south as Colorado and about the lakes at a high elevation, and in the Arctic regions east of the mountains. The nest, composed mainly of feathers and dry leaves, is usually placed in a marsh, and the eggs, of which the number ranges from eight to twelve, are a uniform cream color. When paddling about the marshes, or flying at no height above them, as if seeking some particular spot it could not readily find, this Duck utters a low croaking quack. It feeds upon grasses such as commonly grow in or near ponds and streams, leaves and roots of water plants, and possibly fish, if it can get them, and mollusks; but these last I fancy it eats only when the other more natural food is difficult to obtain.

The male is a very handsome bird, and his stylish, modestly colored dress makes him one of the most attractive of our Water Fowl. There is a good deal of individual variation in the males of this species, and some are more darkly colored than others, and occasionally there is a more or less well defined black ring on the lower part of the neck. The female is a pretty brown and white bird, with a wing somewhat similar to the male's, but without the chestnut on the metallic spot in the center, and by many she is frequently mistaken for the female of the American Widgeon, to which indeed she bears a considerable resemblance.

From its secluded habits the Gadwall is not as well known to the majority of American sportsmen as are the Widgeon and some other fresh-water Ducks, and as it keeps in small flocks and shuns decoys, the opportunities for becoming acquainted with the bird's ways and appearance are at no time very great or favorable. As a bird for the table it is in no way inferior to the Widgeon when both have had access to similar food, and in size the two species are about equal, but if there is any difference the Gadwall may average a trifle larger.

CHAULELASMUS STREPERUS.

Geographical Distribution.—Northern Hemisphere. In North America ranging from Arctic regions to Mexico and Jamaica. Breeds in the Northern States, and in the Arctic Regions east of the mountains.

Adult Male.—Top of head, rufous, varying in depth of shade among individuals, and spotted with black; rest of head light buff or whitish, speckled with blackish brown. Throat, buff, indistinctly spotted with brown. Flesh, dark buff, spotted with blackish. Upper part of back and breast marked with crescent-shaped black and white bars, the former broadest and most prominent. Back, scapulars, and flanks, undulated with slate color and white. Long scapulars, fringed with rusty brown. Lesser wing coverts, gray; middle coverts, bright chestnut; greater coverts, velvety black. Secondaries, pale gray; outer webs, white, forming a speculum beneath the black coverts. Primaries, gray. Crissum and upper tail coverts, jet black. Tail, dark gray, whitish on the edges. Vent and under tail coverts, black; rest of under parts, white. Bill, bluish black. Iris, brown. Legs and feet, orange yellow; webs, dusky. Total length, about 20 inches; wing, $10\frac{3}{4}$; tail, $8\frac{2}{10}$; culmen, $1\frac{8}{10}$; tarsus, $1\frac{7}{10}$.

Adult Female.—Top of head, blackish, faintly marked with buff. Rest of head and neck, yellowish, spotted with blackish brown. Chin and throat, yellowish white, minutely spotted with dark brown. Back and breast, fuscous, the feathers margined with buff. Lower back and rump, fuscous. Wings, like the male, but usually without any chestnut, the wing coverts being gray, tipped with whitish. The speculum is white, with little or no black on its front edge. Primaries, fuscous. Under wing coverts and axillæ, white. Upper tail coverts, fuscous, with V- and U-shaped bars, and edges of buff. Tail, fuscous edged with gray and whitish. Sides, ochraceous, with large spots of fuscous. Anal region and under tail coverts, buff, spotted with fuscous. Rest of under parts, pure white. Bill, dusky, orange near the edges. Legs and feet, dingy yellow; webs, dusky. Smaller in size than the male. Total length, about 19 inches; wing, 10; culmen, $1\frac{7}{10}$; tarsus, $1\frac{4}{10}$.

Young.—No chestnut or black on the wings; white on secondaries not clear; under parts with nebulous brown centers to the feathers. Rest of plumage like the female.

Downy Young.—Forehead and space around the eye, throat, and chest, rich yellow. Upper parts, dark brown, with dark yellow spots on sides of back and rump, and on edges of wing. Lower parts, sooty gray.

EUROPEAN WIDGEON.

A WELL-KNOWN and common species of the Old World, this handsome Duck can only be regarded as a straggler within our limits. It has been killed on numerous occasions in different parts of the United States, usually in the company of the American Widgeon. It is not uncommon among the Aleutian Islands and breeds there, and doubtless individuals starting on the fall migration have taken the wrong course inadvertently, or else have joined flocks of American Wild Fowl and penetrated into unaccustomed lands, and embraced an opportunity to look upon unfamiliar scenes. In its habits it does not vary to any appreciable extent from its American relative, and its life history has been thoroughly written by a number of able English and Continental ornithologists. While having a general resemblance to the Bald-Pate (to anyone who was not accustomed to observe closely), it is in fact a very differently marked bird, and while of very attractive appearance is not as handsome as our own species. Numerous specimens have been obtained in California, and I had a beautiful and very perfect male which was shot in Illinois, and is now with my collection of birds in the Museum of Natural History in New York. I have also seen examples procured on the North Carolina coast, so it would seem that when it strays from its legitimate route, it has no preference as to the road it travels, but visits indiscriminately any portions of the country to which fate may lead it.

27. European Widgeon.

MARECA PENELOPE.

Geographical Distribution.—Northern portions of Eastern Hemisphere, and of frequent occurrence in the United States as far south as California on the Pacific, and the coasts of North Carolina on the Atlantic Ocean. Breeds pretty much throughout the northern part of the Eastern Hemisphere.

Adult Male.—Forehead and crown varying among individuals from white to ochraceous. Remainder of head and sides of neck, rufous. Cluster of small green spots behind the eye, and some on the occiput. Chin, throat, and fore part of neck, black. Breast, vinaceous. Back and sides, finely undulated with black and white. Long tertials, gray on inner webs, black, edged with white, on the outer. Wing coverts, white. Speculum, metallic green. Primaries, fuscous. Lower back, slate gray, with a white spot on each side of base of upper tail coverts. Inner upper tail coverts, gray, with white edges; outer ones, black, edged with white on inner webs. Under parts, white. Under tail coverts, black. Tail, pointed, fuscous, becoming almost black at tips. Bill, bluish black; nail, black. Legs, dark brown; feet, bluish gray; webs, dusky. Total length, 18 inches; wing, 10; culmen, $1\frac{4}{10}$; tarsus, $1\frac{1}{2}$.

Adult Female.—Head and neck, rusty, varying in depth among individuals, speckled with black; occasionally, the top of the head being nearly all black. Upper parts, dusky brown, feathers margined with grayish. Wings, grayish brown. Speculum, dull black; outer web of tertials edged with white, the outermost one with outer web all white. Primaries, fuscous on outer webs, light buff on inner, metallic green at tips. Upper tail coverts, rufous brown. Tail, purplish brown, feathers edged with white. Breast and flanks, light rufous; the former spotted, the latter barred with dark brown. Rest of under parts, white, the under tail coverts barred with blackish brown. Iris, brown. Bill, bluish black. Legs and feet, dark brown. Total length, 18 inches; wing, $10\frac{1}{2}$; culmen, $1\frac{5}{8}$; tarsus, $1\frac{1}{2}$.

WIDGEON.

THE Widgeon is distributed at different seasons of the year throughout North America, from the Arctic Ocean south to Guatemala and Cuba, and from the Atlantic to the Pacific. In its neat dress of attractive colors, some of which exhibit a metallic sheen like burnished metal, the male is one of the handsomest of our Water Fowl, and his demure, modest little consort is no less engaging in her appearance, although not so brilliantly arrayed. The Bald-Pate, another name by which it is known, breeds from the Arctic Sea as far south as the State of Texas, and generally nests in rather high ground in the midst of trees or low bushes, and is not particular about being near water. Its nest is lined with down, and the eggs are covered when the female goes off for any purpose. They are laid in May, and resemble those of the Pintail somewhat, and are a pale buffy white. The male moults while the female is incubating. She takes her turn later. About the latter part of September the young are nearly full grown, and those that have bred in the far North commence their long journey southward. Widgeon are generally observed in small flocks of from six to two dozen, although occasionally great numbers have been seen assembled together, but this is rare. They go much with the Canvas Back and Red Head, and when these dive in the deep water and bring to the surface tender grasses and succulent roots, the Widgeon are very busy indeed stealing these desired objects from their rightful owners, and grow very fat on the fruits of such pilfering habits.

28. Widgeon.

It is essentially a fresh-water species, and keeps to lakes and rivers, and when these freeze resorts to sounds within the beach, where the water may be brackish, or travels on southward to milder climes. The Widgeon is one of the wariest of our Ducks, suspicious of everything, and not only is unwilling to approach any spot or object of which it is afraid, but, by keeping up a continued whistling, alarms all the other Ducks in the vicinity, and consequently renders itself very disagreeable and at times a considerable nuisance to the sportsman. However, its flesh is so tender and palatable, and it is such a pretty and gamy bird, that one is inclined to forgive many of its apparent shortcomings. The usual note of this Duck is a low, soft whistle, very melodious in quality, and when on the wing the members of a flock keep continually talking to each other in this sweet tone as they speed along. They fly very rapidly, and usually high in the air, in a long, outstretched line, all abreast, except, perhaps, the two ends are a little behind the center bird, who may be considered the leader. When only moving from place to place in the marsh, and but a short distance above the ground, they proceed usually without any order or regularity, reminding one sometimes of a flock of pigeons. The pinions are moved with much quickness, and the long primaries give a sharp-pointed shape to the wing that causes the birds to be easily recognized. Flocks composed of a number of Widgeon and Sprigtail are often seen, and the combination is a very unfavorable one to a sportsman who may be hoping for a quiet shot at close range.

As the birds approach the decoys some Widgeon will whistle and edge out to one side, as much as to say, " It may be all right, but I don't like the looks of it," and he will be followed by another suspicious member. Then the Pintails become uneasy and begin to climb and look down into the blind, and the patient watcher sees the flocks too often sheer off to one side and pass him by. But should there be some birds present, as often happens, which are heedless of all warnings or suspicious utterings, and keep steadily on, with the evident intention to settle among their supposed brethren, then, as they gather together preparatory to alight, and the sportsman rises in his ambush, suddenly the air is filled with darting, climbing birds, who shoot off in every direction, but generally upward as if the flock was blown asunder, and all disappear with a celerity that is astonishing, and, to a nervous sportsman, with results that are mortifying.

In various parts of the country this Duck is known by many names other than those already given, some of which are Poacher, Wheat Duck, Bald-Crown and Bald-faced Widgeon, Green-headed Widgeon, Zan-Zan, etc., but among most of the sporting fraternity it is called simply Widgeon or Bald-Pate.

MARECA AMERICANA.

Geogrdphical Distribution.—Throughout North America, from the Arctic Ocean to Guatemala and Cuba. Breeds throughout its range, but chiefly north of the United States.

Adult Male.—Forehead and top of head, white. Behind the eye a lengthened, broad patch of metallic green, extending down hind neck. Rest of head and neck, whitish or buff, thickly speckled with black. Back and scapulars, vinaceous, undulated with black, and, on some feathers, also with white. Wing coverts, white; the greater ones tipped with black, forming a bar across the wing. Secondaries, black, some glossed with metallic green, forming a green and black speculum. Long tertials, pointed, dusky gray on inner web, black edged with white on the outer. Under wing coverts, gray; axillæ, white. Primaries, fawn, shading into glossy brown on outer webs and near tips. Rump and median upper tail coverts, gray, waved with black and edged with white. Outer coverts, black. Breast and sides vinaceous, the inner feathers of the latter undulated with black. Lower breast and abdomen, pure white. Under tail coverts, black. Tail, fuscous, edged with whitish. Bill, pale grayish blue; tip, black. Legs and feet, plumbeous or bluish gray. Webs, dusky. Total length, about 19 inches; wing, $10\frac{1}{2}$; culmen, $1\frac{1}{2}$; tarsus, $1\frac{1}{2}$.

Young Male.—Very similar to the female, but the colors of a deeper and richer hue, the breast and flanks being more vinaceous, and the markings of the wings more clearly defined. The coverts, though dusky in part, have much more white, and the white edges of the tertials are exhibited. The head is much darker as a rule. There is a considerable individual variation seen in this species, especially among adults of, I may say, both sexes. The coloring of the head and neck is frequently different, and this is observable among old males on the top of the head, and in the extent and depth of the green behind the eye, and along the center of the hind neck. The female has some resemblance to that of the Gadwall, but she can generally be distinguished by the coloring of the speculum; the Gadwall's being mostly grayish, while that of the present species is black and green.

Adult Female.—Top of head black, feathers margined with white. Forehead, sides of head, neck, and throat, whitish or buffy white, speckled and streaked with dusky. Upper parts, dusky, barred with buff or ochraceous. Wing coverts, mostly gray, edged with white; the apical half of the outer webs of greater coverts, white, with black tips forming a bar, succeeded by the metallic green and black of the secondaries, making the speculum. Primaries, dusky, fawn color near the shafts. Rump and upper tail coverts, dusky, margined with white. Upper breast and sides, reddish buff or dull vinaceous, the latter barred with dusky, and indistinct dusky blotches on the breast. Rest of under parts, white. Under tail coverts, barred with black and white. Tail, dusky, margined with whitish. Bill, legs, and feet, colored like the male. Total length, about 18 inches; wing, $10\frac{8}{10}$; culmen, $1\frac{4}{10}$; tarsus, $1\frac{4}{10}$.

Downy Young.—Top of head, back of neck, and upper parts, dark olive brown; rest of head and neck, with lower parts fulvous. A dusky streak from bill, through eye to occiput. Spots on posterior border of wing, and on each side of back and rump, greenish buff.

SPRIGTAIL.

THIS is another cosmopolitan species, and ranges in the northern hemisphere, from the Atlantic coast of America across the continent and through the Old World eastward to Japan. It is one of the most common Ducks found in Alaska, and along the mighty river, the Yukon, they nest in May. Mr. Nelson, whose opportunities for watching many species of birds during their breeding season in the Arctic regions have been numerous, describes the peculiar habits of the Pintail at that time. The female rises in the air with the male in quick pursuit, and the pair fly back and forth with incredible speed, performing many quick and varied evolutions, being at one moment almost out of sight overhead and the next just skimming above the ground. The first male would soon be joined by others, all anxious to obtain the fair prize, but none keeps as close to the coy female as the original pursuer, and so dexterous is she in her rapid movements that even he can get near her only occasionally. When he does, however, he keeps beneath her, so closely that their swiftly moving wings rattle together like castanets, the noise thus made being audible for a long distance. This performance is kept up for perhaps half an hour, and all the other males having been distanced in the race, the original pair settle in one of the ponds. At this season this Duck has a habit akin to the drumming of the Snipe. Having risen to a great height, the wings are held stiffly and curved downward, and the bird descends with the swiftness of a meteor, producing

29. Sprigtail.

a sound at first like a low murmur, succeeded by a hiss, and then, as the bird sweeps close along the ground in a gliding course, it assumes almost the proportions of a roar. Sometimes this noise accompanying the Duck's passage through the air is heard for a number of seconds before the bird comes into view, so high has it ascended.

The females of this species are, Mr. Nelson believes, polyandrous, for he has seen one preceded by two males as she flew along, and at short intervals she would halt slightly, draw back her head, and utter a loud nasal quack. It is a common occurrence for a female, when chased by several males, to plunge at full speed under water, followed by her pursuers, and all suddenly rise and take wing a short distance beyond.

The Pintail breeds in northern latitudes of both hemispheres, also in Manitoba and the northern tier of States, occasionally as far south as Colorado, and is among the first of the Water Fowl to commence the duties of incubation; but this important function varies, apparently, according to the degrees of latitude, beginning later in the most northern sections. The nest, composed simply of dry grass and twigs and lined with feathers, is placed in some thick grass, at the foot of a willow, under a bush, or in some similar spot where concealment is equally secured, and from six to twelve pale olive green, rather small eggs are deposited. The young appear during June or early in July, according to the locality, and the parents lead them immediately to the water, from which the nest is never far removed, and they remain about the marshes, keeping themselves well concealed from observation until able to fly. The males moult at this time; the females somewhat later.

In summer the Pintail utters a low, mellow whistle, and also, in addition to the hoarse, guttural quack, a rolling note, similar to that uttered by the little Scaup and some other Ducks, and which can be imitated by a rapid vibration of the tongue, at the same time trying to utter the letter R. The Pintail visits the interior of Alaska as well as the sea-coast, and frequents the pools on the flats. It is also found on various islands of the Aleutian chain. On the eastern side of North America this species is very abundant in summer on the Barren Grounds and in the vicinity of Fort Anderson, where it breeds.

On its autumn migration southward the Pintail makes its appearance in the United States, (provided that it has not remained within our borders during the summer), the latter part of September or beginning of October, coming down with the other "big" Ducks, like the Widgeon, Gadwall, etc., from its northern breeding grounds. On its first arrival it is usually tame and unsuspicious, as there is a large proportion of young birds in the flocks which have yet to make the acquaintance of man and his nefarious ways. At this time they come readily to decoys, and exhibit little of that wariness so conspicuously manifested later in the season.

The lakes and rivers of the western country becoming frozen, the Pintail, in company with the vast army of Water Fowl, now yearly lessened in numbers, moves onward to the milder clime of the Sunny South, or diverges across the country to the shores of the great oceans. On the Atlantic coast multitudes pass the winter in the sounds lying just within the beach of North Carolina and adjacent States, where the usually open winter permits them to indulge in their usual avocations without interruption. Should, however, a cold norther freeze the marshes and open water, they depart temporarily on a brief trip southward, returning again as soon as the weather moderates. By the time they have reached these winter quarters the birds have become "educated," have learned the danger of man's presence, and are generally very shy and suspicious. When coming to decoys, after many hesitating advances, they are apt to rise to a considerable height in the air, and look down into the blind, and not liking the disclosures there made, keep on their course, usually out of gunshot. If they have seen nothing to cause alarm and they come up to the decoys, on the appearance of the sportsman as he rises from his crouching position, the birds seem to throw themselves directly upward at a great speed, with the result of causing any but an experienced gunner to shoot beneath them.

Their flight is very rapid, performed by quick beats of the wings, and the long necks of these Ducks make them easily recognizable when in the air. On the water the Sprig swims gracefully, arching its neck and holding back its head like a miniature Swan, and presents a pretty picture as the sun glances on the variegated coloring of the head and neck of the male. As a diver the Sprigtail is only a partial success. It can go under water, though it cannot stay long, but skulks with great skill, stretching out the neck to the fullest extent and laying it and the head flat upon the surface. At a little distance, unless there is a complete calm, it is very difficult to be seen when it assumes such a position. Beside the names already applied to it in this article, this Duck is known in various parts of our country as Spiketail, Spindletail, Spreettail, Pigeontail, Pian Queue in Louisiana, Water Pheasant, and Smee. Undoubtedly it has other local names besides these.

DAFILA ACUTA.

Geographical Distribution.—Cosmopolitan. In North America it ranges from Alaska to Panama and Cuba. Breeds from northern United States to limit of its northern range.

Adult Male.—Head and upper neck, hair brown, darkest on the crown, where it is often a rusty brown. Sides of occiput with metallic green and purple reflections. Upper part of hind neck, black; lower part, dusky, minutely waved with white. A white stripe, beginning at the upper edge of black portion, passes down the sides of the neck, and is confluent with the white of the under parts. Back, and sides of flank, waved with narrow white and dusky lines. Tertials, silvery gray, with a central black stripe; long scapulars, black, edged with buff or whitish. Wing coverts, glossy brownish gray, last row tipped with cinnamon, forming a bar across the wing. Speculum, bronze, changing from green to copper according to the light, with a subterminal black bar and white tip. Under parts, pure white, sometimes blotched with rust color. Sides and flanks crossed with narrow bars of white and dusky. Lengthened upper tail coverts, black, edged with white on inner webs. Tail feathers, pointed, dark brown on outer webs, gray on inner, the long central pair narrow and pointed and extending beyond the others, black. Under tail coverts, black, the external ones having white outer webs, forming a line on each side. Iris, dark brown. Bill bluish gray, blackish toward tip; lead color toward the edges. Legs and feet, brownish gray. Length, about 26 inches; wing, $10\frac{1}{2}$; culmen, $2\frac{8}{10}$; tarsus, $1\frac{6}{10}$; tail, 7; bill, 2.

Adult Female.—Top of head, rufous streaked with black. Rest of head, whitish or yellowish white, finely streaked with dusky. Back of neck, dusky, streaked with buff; chin and throat, whitish; upper parts, dusky, crossed with irregular, often U-shaped, bars of yellowish white, or ochraceous, these last being mostly on middle of back. Wing coverts, brownish gray tipped with white. Under parts, white, streaked with dusky. Sides and flanks with broad V- or U-shaped marks of glossy grayish brown. Upper tail coverts, irregularly blotched with black and white. Tail, dark brownish gray irregularly barred with white. Bill, bluish gray, blackish on top. Legs and feet, lead color. Length, about 20 inches; wing, $9\frac{8}{10}$; culmen, $1\frac{8}{10}$; tarsus, $1\frac{6}{10}$.

Adult Male in Moulting Plumage.—Like the adult female, but darker, and exhibiting a brilliant speculum.

Young.—Also like the female, the males always distinguishable from the females by having a speculum on the wing.

Downy Young.—Crown of head, back of neck, and upper parts, olive brown, with a dull white stripe on each side of back. Yellowish white stripe over eye, and a brown one through the eye from bill, and a spot of the same color over the ears. Lores, brownish. Under parts, grayish white.

BLUE-WINGED TEAL.

MORE restricted in its range than the Green-Winged Teal, the present species, sometimes called Summer Teal and White-faced Duck (Printempsnierre in the spring, and Automnierre in the autumn in Louisiana), is found chiefly in the Mississippi Valley, where it is very abundant, and throughout the eastern portion of the United States. It is rare in Alaska, and is accidental on the Pacific coast north of the Gulf of California, save, perhaps, in summer, when it occasionally appears upon the Alaskan coast. In winter it goes south to the West Indies and northern South America. The Blue-winged Teal breeds in various portions of the eastern States of the Union, and also in the Mississippi Valley, and is one of the first of the great host of the Duck tribe to appear in the annual migration southward.

This Teal nests on low land, usually near the water, amid reeds and high grass growing in such situations. In the center of a mass of rushes and coarse grass a quantity of down and feathers is placed, and upon this sometimes as many as twelve white eggs are deposited. This Duck is a lover of mild climates and soft airs, and is never seen when ice and snow abound, unless some such calamity as a severe frost has suddenly come upon the southern land in which it is passing the winter. Early in September the flocks gather in the northern part of the Union, preparatory to their departure on their southern journey, while those which have passed the summer north of our borders commence to appear within the

30. Blue-Winged Teal.

United States. They come in large flocks, and frequent the inland lakes and rivers, feeding upon insects and tender plants and grasses. Wherever the wild rice grows, there, in autumn, are these Teal found, and they scatter themselves throughout the matted growth of this plant, which frequently spreads over a large portion of the bottoms of many of our western lakes and rivers. Here the Teal are safe so long as they remain in the interior of the beds, for nothing of the earth or air can reach them as they paddle about hidden in the deep recesses of the wild rice. They feed upon the ripened grains that fall upon the water, or dig them out of the mud upon the bottom, and become exceedingly fat. I know no better bird for the table than a Blue-winged Teal fattened upon wild rice. Many are killed by sportsmen stationing themselves just within the borders of these rice beds, and shooting the birds as they fly over or around, looking for a favorable place to settle. When feeding the members of a flock keep as near together as possible, and rarely utter any sound, each one too intent apparently upon his own affairs to indulge in the pleasures of conversation. When startled it rises from the water by a single spring, and the flight is exceedingly rapid, and it has the habit of turning alternately to the observer the upper and lower surface of the body as it speeds along, rolling, in fact, in a similar manner to a boat in the trough of the waves. This species utters at times a lisping note when on the wing, and should it perceive a desirable place for feeding, or a number of its fellows congregated together, it drops suddenly into the water, without making any elaborate preparation to alight, but simply stops at once. I have never found it a shy bird, for it usually allowed me to approach closely without showing any especial alarm, and it always came boldly in to the decoys, and, if permitted, settled among them in full confidence, and began to swim about its supposed brethren quite at home and contented. When on the water this Teal swims with much buoyancy, and the flocks, like those of the Green-winged Teal, are compacted together so closely that the members would seem to be in each other's way as they floated along. In February the movement toward northern climes begins, and, like all Ducks at this season, they are poor in flesh and should never be shot. The male has assumed the summer dress, one of the most beautiful among the Duck tribe, and the pure white crescent before the eye makes him very conspicuous as he paddles about the ponds and inlets, or wanders over the muddy bars in quest of food. Like its relatives, the Blue-winged Teal walks easily and well, and is able even to run quite rapidly.

QUERQUEDULA DISCORS.

Geographical Distribution.—North America, but chiefly in the eastern portion; Alaska, and south to the West Indies, and northern South America. Occasional in California. Breeds from Kansas northward.

Adult Male.—Top of head, black, feathers edged with ochraceous. Chin and space along base of bill, black. A large crescent-shaped band, white edged with black, goes from the forehead in front of the eye to the throat. Rest of head and neck, dull plumbeous, with a metallic purple gloss on the occiput. Back, dusky, with U-shaped bars of buff. Long scapulars, greenish black, with a central stripe of buff. Lesser wing coverts and outer webs of some scapulars, pale blue. Greater coverts, dusky, with white tips forming a bar in front of the speculum, which is metallic grass green. Lower back, upper tail coverts, and tail, dusky, feathers of the last two margined with whitish. A white patch on each side of the tail. Entire under parts and sides, reddish buff, inclined to pale chestnut on lower breast. Under tail coverts, black. Bill, black. Iris, brown. Legs and feet, yellow, with the web dusky. Total length, 15 inches; wing, $7\frac{8}{10}$; culmen, $1\frac{7}{10}$; tarsus, $1\frac{9}{10}$.

Adult Female.—Top of head, black, remainder of head and neck, brownish white, speckled or streaked with dusky. Chin, throat, and base of bill, white. Upper parts, dusky, barred with V-shaped buff marks. Wing coverts, blue, like the male, but the green speculum is wanting. Upper tail coverts and tail like the male. Under parts, pinkish buff on the breast, with dusky V-shaped marks, remaining portion white indistinctly spotted with dusky, most numerous on the under tail coverts. Bill, greenish black. Legs and feet, pale flesh color. Total length, 15 inches; wing, 7; culmen, $1\frac{4}{10}$; tarsus, $1\frac{1}{2}$.

Young Male.—Similar to female on head, neck, and upper parts. The white throat is speckled with dusky. The green speculum is visible, and the under parts are like the adult male, with the flank feathers broadly barred with dusky.

Young.—Like adult female, but with a pure white belly and grayish brown speculum.

CINNAMON TEAL.

THIS rather handsome bird is restricted to the western portion of North America, from the Columbia River, along the Pacific coast, south to Chili, and eastward to the Argentine Republic and the Falkland Islands. Occasionally it straggles into the Mississippi Valley, and has even been known to go as far eastward as Florida, but such occurrences are extremely rare and can only be regarded in the same light as would be the appearance of some European species taken within our limits. In the United States the Cinnamon Teal is essentially a western bird, particularly numerous in California, where it is found in flocks of considerable size, and associates with other fresh-water Ducks. It goes in summer as far north as the upper part of the Columbia River, and has been found nesting in Idaho, and breeds in various parts of Colorado. It is abundant also in the great Salt Lake Valley.

The breeding season commences in May, about the middle of the month. The nest is composed of grass, lined with down and feathers, and placed upon the ground, generally in the vicinity of water, and about a dozen creamy-white eggs are deposited. In its habits this species does not differ appreciably from its eastern ally, the Blue-winged Teal. It flies as swiftly, rises as suddenly from the water when startled, and is as palatable as an article of food. It would seem that South America was more naturally its home, and its dispersion is greatest on that continent, and that the western section of our

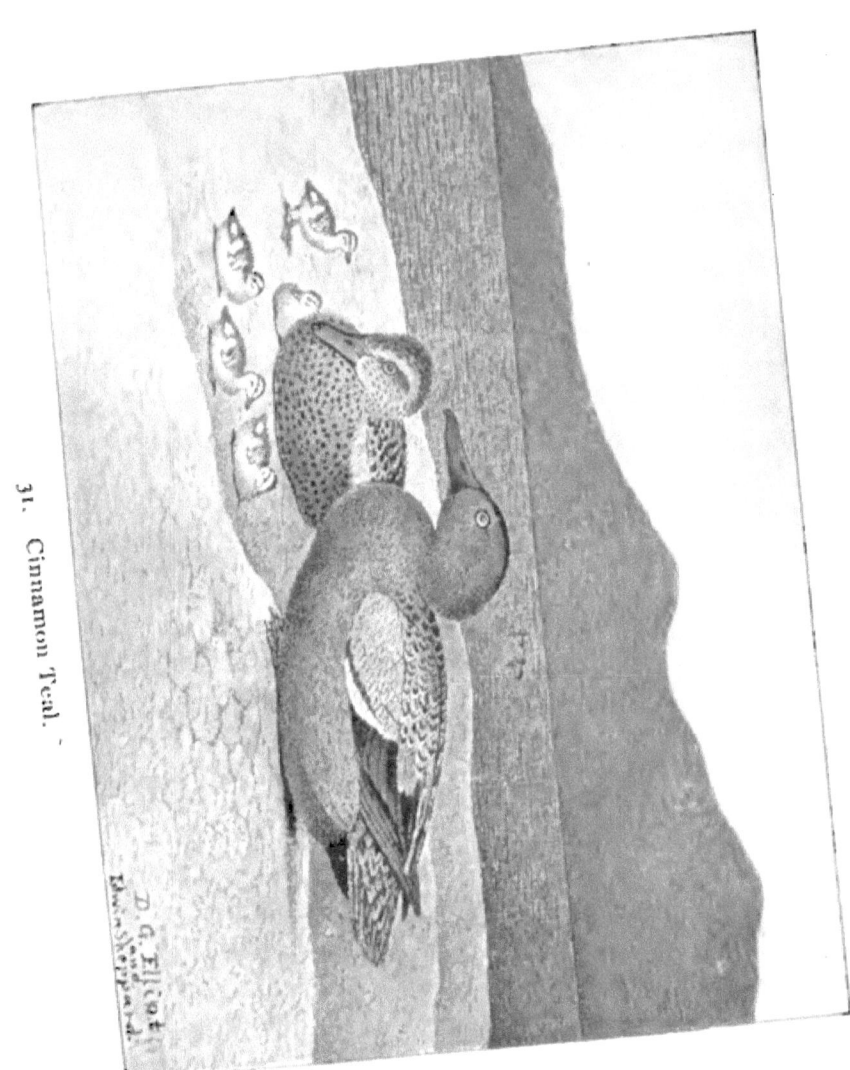

31. Cinnamon Teal.

own land was but an outlying district of its true habitat. The male is a handsome bird in his purplish chestnut dress.

QUERQUEDULA CYANOPTERA.

Geographical Distribution.—Western America, from British Columbia south to Chili, Patagonia, and the Falkland Islands; east to southern Texas; casual in the valley of the Mississippi, and certain of the eastern States as far as Florida.

Adult Male.—Top of head, blackish chestnut. Rest of head, neck, and lower parts, uniform bright chestnut. Back, rump, upper tail coverts, and tail, fuscous with light edges. Scapulars, chestnut barred with black, the long ones, black with a buff central stripe, and pointed. Wing coverts and outer webs of some scapulars, pale blue. Tips of greater wing coverts, white, making a bar above the bronzy green speculum. Under tail coverts, blackish. Bill, black. Legs and feet, orange; webs, dusky. Iris, orange. Total length, about 17 inches; wing, $7\frac{1}{4}$; culmen, $1\frac{8}{10}$; tarsus, $1\frac{1}{4}$.

Adult Female.—Similar to the female of the Blue-winged Teal, but more reddish. The sides of head and throat, deep buff, and the back, fuscous, the feathers edged with pale buff as in *Q. discors*. The entire under parts are light brown, inclining to rufous on upper breast, which is spotted with black or dusky; rest of under parts indistinctly barred with chestnut on abdomen, and with nebulous dusky spots on anal region and under tail coverts. Wings as in female *Q. discors*, but with a speculum faintly defined, of dark green. Bill, dusky, pale on the edges. Iris, brown. Feet, yellowish drab. Total length, about $16\frac{1}{2}$ inches; wing, $6\frac{8}{10}$; culmen, $1\frac{1}{10}$; tarsus, $1\frac{7}{10}$.

Young Male.—Like female, but under parts streaked instead of spotted.

Downy Young.—Top of head, hind neck, and upper parts, olivaceous, darkest on the head; forehead, stripe over the eye, sides of head and lower parts, yellowish buff. A narrow dark brown stripe on sides of head, greenish buff spots on sides of back, and yellowish spots on sides of rump.

EUROPEAN TEAL.

THIS well-known resident of the northern portions of the Old World bears a very close resemblance to the Green-winged Teal of our own land. It is only a straggler within our limits, individuals having been taken occasionally on the northern part of the Atlantic coast, waifs probably from Greenland, where it is sometimes found, which have wandered down our shores instead of taking their legitimate route to the Eastern Hemisphere. The European Teal also occurs at times in the Aleutian Islands, and Mr. Turner procured a specimen on Atkha. It is probably a summer visitant to that chain of islands, and may breed there. Although I have never met with this species alive in North America, I have frequently seen it in the markets of New York hanging with other ducks procured along the shores of Long Island and other near points upon the coast. While possessing a number of differences more or less pronounced from the American species, it is mainly recognizable by the absence of the conspicuous white bar on each side of the breast, which is an especial feature in the plumage of our Green-winged Teal. In the Old World this Teal is generally distributed from the British Islands to China and Japan. It can be domesticated without difficulty, bears confinement well, and breeds readily if suitable locations are provided for it. It is a very pretty species, and does not differ in economy and habits from our own bird.

32. European Teal.

NETTION CRECCA.

Geographical Distribution.—Northern portions of the Old World. Occasional in North America.

Adult Male.—Very similar in plumage to the American Green-winged Teal, but with the following differences: Green band behind the eye, bordered anteriorly with yellowish white, more conspicuous than in the American species; there is no white bar in front of the bend of the wing. The black and white undulations on back and sides are much coarser; the outer scapulars have the inner webs entirely, and the outer partly, white, or yellowish white, while the exposed portions of outer webs are black, forming two broad stripes down the wing, the inner white, outer black. The remainder of the plumage is practically indistinguishable from *N. carolinensis*, the American species. Bill, black. Legs and feet, brownish gray. Total length, 14 inches; wing, 7; culmen, $1\frac{1}{2}$; tarsus, $1\frac{1}{4}$.

Adult Female.—Very like the same sex in the American Green-winged Teal, so much so that anyone might be excused for confounding them. The back is fuscous, but the bars and margins of the feathers are throughout of a deeper hue, more generally ochraceous than buff. The sides of the head, neck, and throat are deep buff, much darker than the same parts in its American ally. These seem to be the only tangible differences in the specimens before me, and they may be to a great extent individual, and the only way that a specimen of a female can be determined with any certainty is to have the locality in which it was procured established without doubt. Even then, in the case of a female of the European Green-winged Teal, killed in America, it would be a difficult task to decide as to which species it belonged. Total length, 13 inches; wing, $6\frac{8}{10}$; culmen, $1\frac{4}{10}$; tarsus, $1\frac{1}{4}$.

Downy Young.—Line on forehead, top of head, back of neck, stripe through eye to occiput, and one from corners of mouth to and including ear coverts, and entire upper parts, dark brown. Sides of head, buff; throat and under parts, and spots on shoulder, and on each side of back and rump, yellowish white. Bill, black; tip, orange.

GREEN-WINGED TEAL.

A BEAUTIFUL bird, the American Green-winged Teal has a very extended distribution in North America, and ranges from the Arctic Sea across the entire Continent from the Atlantic to the Pacific Ocean, and south to Honduras, in Central America, and to Cuba. It breeds as far south as Colorado, but goes mostly north of the United States for the purpose of incubation, and is very common in summer in Alaska and among the islands of the Aleutian chain, and also on the eastern portions of the continent, in the valley of the Saskatchewan, the Mackenzie River district, and about Hudson Bay. It makes its nest in tall grass or in clumps of dried grass and feathers, and lays from eight to a dozen ivory white eggs. Incubation commences the last of May, and the young are hatched by July. This species goes in large flocks, and flies with great swiftness, at times keeping a straight course, as though its destination was unalterably fixed in its mind and it intended to reach it by the shortest possible route, and again it will be irregular and vacillating in its movements, changing its course frequently and dodging about with as much eccentricity of action as that exhibited by a butterfly in a strong breeze. But whatever may be its movements, its flight is always rapid, and its small body proves to be an exceedingly difficult mark to hit.

Although usually breeding north of the boundary between the United States and Canada, it has been known to nest in Wisconsin, Iowa, and others of the northern

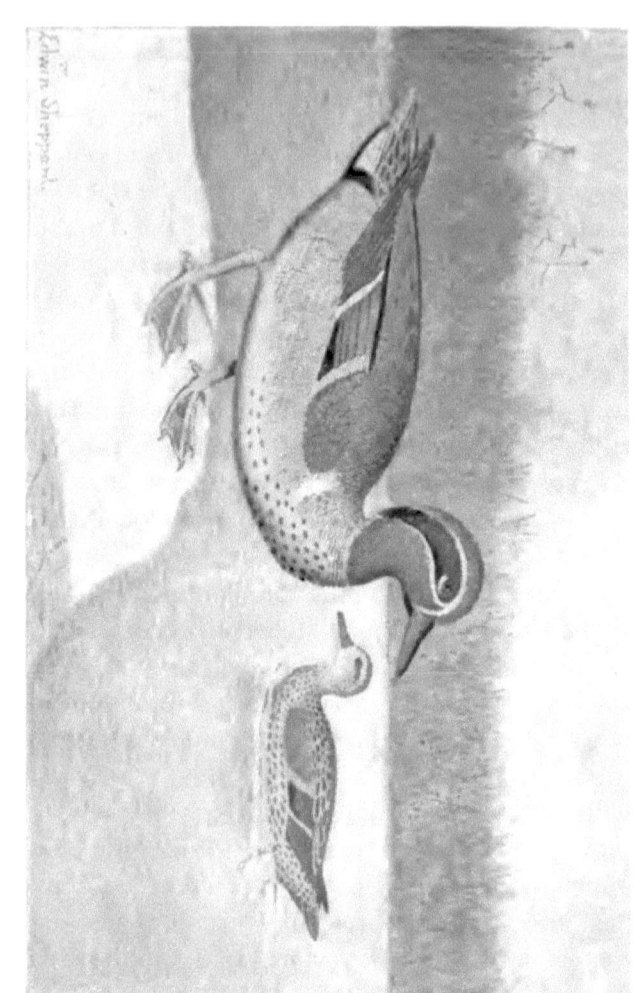

33. Green-Winged Teal.

tier of States, and in some localities seems to prefer the neighborhood of small streams to the larger bodies of water equally available. Occasionally very large broods are seen, whether the product of one female or from two having used the same nest it was impossible to determine, but Hearne states that at Hudson Bay he had seen the parents swimming at the head of seventeen young, and that the latter were not bigger than walnuts. No wonder that the species is able to keep up its numbers fairly well, even against the immense adverse interests that hasten its destruction, when it can claim among its members such patriotic and prolific parents as those above mentioned.

The Green-winged Teal is a fresh-water bird, and although it visits the sea-coast, it keeps to the marshes and tidal creeks and rivers. The flocks swim closely together, rarely scattering about much even when feeding (at least that is the way they generally acted when I observed them), and were very quick in all their movements, sitting, if not alarmed, rather high on the water. It is an expert diver and can remain beneath the surface for a considerable time. It rises with a sudden spring and is at once in full flight, and it requires a marksman with a steady eye and hand to make a successful shot at one of these birds on the wing. It passes southward from its northern breeding grounds in October, being somewhat later in its migration than its near relative, the Blue-winged Teal, and visits the ponds, small lakes, and streams, feeding on insects and various leaves and grasses. In the South it visits the rice-fields, and keeps company with Mallards and other large Ducks found in such places. Like all Water Fowl, this Teal feeds much at night, particularly if the moon is shining, but if in localities where it is not much disturbed, it is also very active during the day. After feeding I have often seen large flocks gather on a lake or broad place on a river, notably the Mississippi, and huddled closely together enjoy a quiet siesta in the warm sunshine, and, in the case of the river above mentioned, floating along with the current, apparently utterly indifferent as to where it might carry them. The flesh of this Duck is very tender and of excellent flavor, especially when the bird has been feeding on delicate grasses, like the wild celery or similar food, and on this account is much sought after by gunners. It is, however, of better flavor when procured in the interior than on the seacoast, its food in the latter locality probably being of a less desirable quality. In addition to the name at the head of this article, this species is also called Mud Teal, Winter Teal, Red-headed Teal, and Sarcelle by the French.

NETTION CAROLINENSIS.

Geographical Distribution.—Throughout North America from the Arctic Regions to Honduras in Central America, and to Cuba. Breeds north of the United States, only occasionally within our limits.

Adult Male.—Head and neck, rufous chestnut, with a broad metallic green band from eye to nape, terminating in a tuft of purplish black. A narrow buff line borders the under side of the green band. Chin, black. Back and sides crossed with narrow, wavy black and white lines; lower back, dark brownish gray. Upper tail coverts, dusky; margined with white. Tail feathers, brownish gray, edged with white. A broad white bar in front of bend of wing. Wing coverts, brownish gray, tipped with ochraceous buff, forming a half bar across wing, succeeded by a broad metallic green patch or speculum, bordered beneath by another broad black bar, tipped with white. Tertials, brownish gray on inner webs, crossed by narrow black and white wavy lines on outer webs, and margined with black. Secondaries, brownish gray; the outer ones bordered with black, which with the same color of the tertials forms a narrow stripe along the wing above the coverts and speculum. Primaries, brownish gray. Breast vinaceous, covered with round black spots, growing indistinct on the lower part of breast. Belly, white; sometimes tinged all over with buff. Buff patch on either side of crissum. Under tail coverts, black. Bill, black. Legs and feet, bluish gray. Total length, $14\frac{1}{2}$ inches; wing, $7\frac{1}{4}$; bill, $1\frac{1}{2}$; tarsus, $1\frac{1}{4}$.

Adult Female.—Top of head and hind neck, fuscous; feathers margined with ochraceous. Sides of head and neck, buffy white, speckled with dusky. Chin and throat, buff. Upper parts, dusky; feathers, barred and margined with pale buff and ochraceous, intermixed. Wing similar to the male, the speculum smaller, and the tertials colored like the back. Rump and upper tail coverts, fuscous, margined with white. Tail, pointed, fuscous, edged with white. Upper part of breast, dark buff, spotted with fuscous. Rest of under parts, white, with nebulous dusky spots, most numerous on anal region and under tail coverts. Bill, black. Legs and feet, bluish gray. Total length, $14\frac{1}{4}$ inches; wing, $6\frac{7}{10}$; culmen, $1\frac{1}{10}$; tarsus, 1.

Young Male.—Sides and belly, pure white; rest of plumage like female.

Downy Young.—Head, neck, and lower parts, pale buff; darkest on top of head and nape, which is grayish brown. A dusky stripe behind the eye, and a dusky spot over the ears. Upper parts, grayish brown, with a buff spot on sides of back and rump.

SHOVELER.

A THOROUGHLY cosmopolitan species, the Shoveler, or Spoonbill as it is often called, is found pretty much everywhere throughout the Northern Hemisphere, and may penetrate possibly into the limits of the Southern also, although there, in different parts, it is replaced by other species of the genus. In North America it is generally distributed, but is not common on the eastern coast, and breeds from Alaska to Texas. It is a fresh-water Duck, and is fond of resorting to inland lakes and streams, and seeks places overgrown with plants and rushes, feeding on seeds, insects, and such food as it is able successfully to sift through its heavily fringed bill, being more generously provided in this respect than almost any other Duck. The Shoveler is usually seen in flocks, some of considerable size, and, when in the air, its long, sharp-pointed wings with their wide expanse give the bird the appearance of being much larger than it really is. In Alaska, as would naturally be expected, the Shoveler is not common along the sea-coast, but breeds in the interior, and is rather abundant in certain portions of the Yukon. It has also been met with on the Commander Islands, and in Kamchatka. When about the marshes, or moving over the inland lakes and coasting along the shores, the Spoonbill is readily recognized by its flight, which is more like that of a Teal, although much less swift, and is performed in an irregular, hesitating kind of way, as if the bird was uncertain just where to go, and it moves in and out among coves and

34. Shoveler.

creeks, apparently investigating every spot, as if searching for some specially suitable place to alight. It is not particularly timid, and will often come boldly up to decoys, looking really quite like one of the " big " Ducks as it sets its wings and sails up to the wooden counterfeits. But in reality the body of the Shoveler is not large, and its apparent size, in the air, is mainly made up of wings and head, of which the huge spoon-shaped bill is not the least portion. It breeds early in the year, the month largely dependent upon the latitude in which the bird happens to be, as there is great diversity of climate between the limit of its northern and southern dispersion, and it is apparently a species that breeds wherever the proper season of the year for that duty happens to find it.

The nest, composed of grass or rushes laid upon a dry spot on some low land near water, is lined with feathers from the parent's breast, and from eight to a dozen greenish white eggs are laid. The young have a bill shaped like that of any other Duck, the broad overlapping maxilla not being developed until the bird is well grown. The male Shoveler in full summer dress is a very handsome Duck, indeed, of particularly striking appearance; its dark green head and neck, somewhat like the Mallard's, showing with much effect above the white breast, and both finely contrasted with the deep chestnut of the under parts. It is not a graceful bird, its huge bill giving it a topheavy look, but it walks well on land, and can run with some speed. I have seldom heard the Spoonbill utter any sound, though occasionally it gives forth a few feeble quacks, but it is usually very silent. As a bird for the table I have held it in very high esteem, its flavor depending greatly, of course, on the quality of food it obtains. This species has many local names by which it is known to sportsmen and gunners. Some of these are, Spoonbill, Blue-winged Shoveler, Red-breasted Shoveler, Spoonbilled Teal, Spoonbilled Widgeon, Broad Bill, Broady, Swaddlebill, Mud Shoveler, and in Louisiana, Mesquin.

SPATULA CLYPEATA.

Geographical Distribution. — Cosmopolitan. Throughout the Northern Hemisphere. In North America from Alaska to Texas, and thence southward through Mexico and Central America to northern South America. Not common on the Atlantic coast. Breeds pretty much throughout its range.

Adult Male.—Head and neck, dark metallic green; black in certain lights. Upper part of back, breast, and anterior scapulars, white. Middle of back, brown; rump and upper tail coverts, black, glossed with metallic green. Wing coverts and outer web of two long scapulars, pale blue. Tips of greater wing coverts, white, forming a narrow band across the wing. Speculum, metallic grass green. Inner secondaries, greenish black, with median white stripe. Primaries, fuscous on outer webs, pale brown on inner. Tail, with central rectrices, brown, edged with white; remaining feathers, white; freckled or blotched with brownish gray. A white patch on each side of root of tail. Entire under parts, rich deep chestnut, extending to crissum, which with the under tail coverts, is dark metallic green, black in some lights, bordered anteriorly by a narrow band of white, undulated with black. Inner feathers of the flanks, pale chestnut, freckled with black. Bill, black; iris, pale yellow. Legs and feet, orange red; webs, violet gray. Total length, about 19 inches; wing, $9\frac{1}{2}$; culmen, $2\frac{7}{10}$; tarsus, $1\frac{4}{10}$.

Adult Female.—Front and top of head, brownish white, streaked with dusky; neck and sides of head buff, streaked with dusky. Chin and throat, uniform buff. Upper part of back and wings, fuscous; feathers, edged with yellowish white. Wing coverts, dull, pale blue; feathers sometimes edged with white. Speculum, metallic green. Middle of back and rump, fuscous; feathers, edged with V-shaped bars of reddish buff. Upper tail coverts, fuscous; irregularly barred with buff or white. Tail, white, barred with brown. Under parts, reddish buff, spotted with brown. The abdomen sometimes immaculate white. Bill, olive brown, sometimes speckled with black; base of maxilla and all of mandible, orange. Iris, yellow. Legs and feet, orange. Total length, about 19 inches; wing, $8\frac{3}{4}$; culmen, $2\frac{1}{2}$; tarsus, $1\frac{3}{10}$.

Young Male.—Resembles the female, but has the head and neck mottled with black, and the black feathers on top of the head are edged with reddish buff. The upper part of breast and back is pale reddish buff with V-shaped marks of dark brown. Rest of upper parts like the female. The under parts, pale chestnut; but there is much individual variation in the coloring of lower breast and abdomen. Wing very like that of the adult male.

The male, in full breeding plumage, is not commonly met with; but this species, in all its variety of dress, with the exception of the Downy Young, is readily recognizable by the peculiarly shaped bill.

Young Female.—Speculum, dusky, with little or no metallic reflections, and tipped with brownish white. Wing coverts, slate color.

Adult Male, when moulting, resembles the female, but is darker, and the speculum more brilliant.

Downy Young.—Middle of crown, nape, and hind neck, olive brown; rest of head and neck, and lower parts, pale fulvous. A dark brown stripe from bill through eye halfway to occiput, and a similar one across ears toward nape. Upper parts, olive brown, with yellowish spots on each side of back and rump.

RUFOUS-CRESTED DUCK.

THIS is a species of the Old World, and is very questionably included in the North American Fauna. So seldom has it been obtained within our limits that it can hardly be considered even as a straggler; the few specimens known having been seen hanging in the market in New York for sale, but the locality from whence they came was very doubtful, and it was only the fact that the birds were in the flesh which gave rise to the thought that they might have been killed within our boundaries. Many European game birds are exhibited for sale in our Eastern markets during the winter that were never killed on our shores, as invoices of them are brought by nearly every steamer, and it is only because it would be considered doubtful that anyone should send a Wild Duck to America, it being an act very near akin to shipping coals to Newcastle, that it became a fair supposition that these specimens of this Duck came to our shores by means of their own propelling powers, unassisted by man.

The Rufous-crested Duck is a very handsome species and in the Old World is found in southern and eastern Europe, occasionally straggling into the northern parts of central Europe, and also is an inhabitant of Northern Africa and India. It frequents often fresh-water lakes and marshes, and is very shy, and has a note resembling the harsh croak of the crow. It is not a diver, and feeds from the bottom, like the Mallard, by tilting its hindquarters, and holding itself in position by paddling with

35. Rufous-Crested Duck.

the feet, while it pulls up the grass and plants growing below. It goes in small companies and does not consort with other species. It breeds in ponds, the nest being placed amid rushes or flags, and is composed of these plants, dead leaves, and a bed of down. The eggs, which are an olive-green, vary from eight to ten. While incubation proceeds, the males assemble together on the water in the vicinity. Whenever the female leaves the nest, she covers the eggs carefully with down. In Italy this is a common species, and also in India, where it keeps to the middle of the tanks, and is very wary and difficult to approach. Its flesh is considered excellent, and by some regarded as one of the best birds for the table found in that country.

With all its favorable qualities, both of attractive appearance and palatable flesh, it is to be regretted that this Duck can in no wise be enrolled in our lists as belonging to North America. It is one that would be much better dropped from our catalogues as an American species, and erased, with some others of equally questionable standing, from our scientific works. It is included in this book simply because it has been retained in the Check List of the American Ornithologists' Union, as it seems best to me not to omit any species given in that catalogue.

NETTA RUFINA.

Geographical Distribution.—Eastern hemisphere. Of questionable occurrence in eastern United States.

Adult Male.—Sides of head and throat, vinaceous, darkest on the throat, passing into pale rufous on the front and base of crest, grading into pale reddish buff on the central portion of the latter. Upper part of back of neck, and all lower neck, black, grading into the glossy blackish brown of the breast, belly, and under tail coverts. Upper back, grayish brown, passing into chocolate brown on the rump; upper tail coverts, black, with a greenish gloss. Scapulars, yellowish brown. Joint of wing, and a patch continuous with it, partly concealed by the scapulars, white. Wing-coverts and tertials, grayish brown; secondaries, white tipped with gray forming the speculum. Primaries, white, the tips and outer web of the first five dark grayish brown. Sides and flanks, white suffused with pink undulated with dark brown bars anteriorly and posteriorly, some indistinct. Upper portion of flanks bordered with reddish brown. Tail, grayish brown, pale on inner webs. Bill, vermilion red. Iris, reddish brown. Legs and toes, vermilion red; webs, blackish. Total length, about 22 inches; wing, 10; tail, 4; culmen, 2; tarsus, $1\frac{6}{10}$.

Adult Female.—Upper part of head, dark brown; back of neck, pale grayish brown; cheeks, throat, and sides of neck, grayish white. Entire under parts, brownish white, passing into pure white on the under tail coverts. Upper parts, grayish brown, grading into blackish brown on the rump. Scapulars, grayish brown, paler than in the male. Wing coverts, pale grayish brown. Secondaries, white, forming the speculum. Primaries, grayish white; outer webs and tips, dark brown. Upper tail coverts, pale grayish brown. Culmen, $1\frac{2}{3}$ inch; wing, 10; tail, $3\frac{2}{3}$; tarsus, $1\frac{3}{4}$.

Downy Young.—Superciliary stripe, and one through the eye dividing into two posteriorly, buff. Upper parts, olive gray. Spot on each shoulder, and entire under parts, buff.

36. Canvas Back.

CANVAS BACK.

GIVEN the proper kind of food, there is no Duck, save perhaps occasionally the Red Head, that can equal this splendid species in the delicate quality and flavor of its flesh, and as a game bird and for the sport it affords to the gunner, there is no Water Fowl worthy of being mentioned with this one, so deservedly known as the Royal Canvas Back. Exclusively an American species, having nothing in the Old World that can even be said to represent it, the Canvas Back ranges over all North America, and breeds from upper California, amid the lakes and water courses of the mountains, in eastern Oregon in similar lofty situations, and in some other States on our northern border, to and throughout the Arctic regions probably to the sea. It is not found, however, on the Pacific coast north of Vancouver Island. At different points on the Yukon it breeds in great numbers, and probably its main nesting ground is in that northern latitude. The places within the limits of the United States suitable for this Duck to rear its young unmolested will probably grow fewer and fewer, until in a brief period it will have to rely altogether upon Arctic solitudes for that protection and freedom from intrusion so necessary at this important period of the bird's existence. The bottom of the Canvas Back's nest is formed of rushes and grasses situated in the water, and is then built up with high sides and lined with down and feathers. It is continually being added to while the bird is laying, and when the female is ready to commence incubating, it has grown to be considerable of a structure. Eight to ten pale greenish gray eggs are deposited, and the female begins to lay about June.

The Canvas Back appears within the limits of the United States, during the fall migration, in the month of October. The duties and trials of the nesting season and the rearing of the young broods in the far northern regions are over, and each little family, lusty of wing and robed in a fresh dress, has united itself with some others until the gathering host, making ready for the long southern journey, spreads itself out like some great army preparing to invade an unknown country. The sun has for some time been making his daily rounds in constantly diminishing circles, and the increasing time between his setting and rising, with the gradual lengthening of the period of darkness, all betoken the coming of the Arctic night. It is time for birds to be on the wing, headed for southern climes. Preparations are made for their departure and much discussion must be indulged in, probably both as to what they expect to see and find in this, to many, *terra incognita*, and as to the best routes to reach it. Some are present who have made the journey, perhaps many times before; wise old heads that have escaped unnumbered dangers and traps set for the unwary, and who have sturdily refused to listen to the charm of the sportsman's well-imitated call,—charm he never so wisely,—or to be allured into the dangerous neighborhood of his ambush, be his decoys ever so lifelike and competent to deceive. But the majority of that preparing host are young and inexperienced, ignorant of all that is before them, and of the dangerous ways of the world. But they must take their chances, like all the rest of earth's creatures in the great struggle for existence, and the time has come to depart.

With a roar of wings like the sound of many waters, as if actuated by a single impulse, the feathered army rises in the air, and captained by a few old birds, survivors of many a battle, the return journey commences. With a few preparatory wheels around the vicinity of their summer home, which many of them will never see again, the leaders head to the south, and, at a lofty height, guide the main body at a great speed toward the promised land.

On Puckaway Lake, in Wisconsin, Canvas Backs and Red Heads would always make their appearance on the 10th day of October. It was a very singular fact, but we could always be certain of seeing some of these Ducks at that date; no matter what the weather may have been up to that time, and even if the season had been unusually cold, these birds did not appear before the 10th. The lake contained plenty of wild rice and celery, and before it was closed by ice the Canvas Back would become very fat upon this food, and were not surpassed in delicacy of flavor by any shot upon the famed waters of the Chesapeake. Like the Red Heads and some other diving ducks, the Canvas Back keep out in deep water and raft together in great numbers, seeking their food at the bottom. Their feet, although large and powerful, are not of much assistance in descending to the depths, but the wings are the bird's chief reliance for propulsion, and it flies under water as it does in the air, and the feet are employed mainly for guiding and altering the course. This method of propelling itself under water is not by any means the sole attribute of the Canvas Backs, for not only do many other Ducks act in the same way, but different species of water birds, not Ducks, also.

The flight of the Canvas Back is not probably exceeded in swiftness by that of any other Duck, and under favorable circumstances it will doubtless accomplish one hundred miles an hour. It generally flies in a direct line as if it knew exactly where it was going, and often at a great height. Its method of flying resembles very closely that of the Red Head, and it moves along in extended lines in the way described in the article on that bird. It is also in the habit of exercising in the early morning and late afternoons. The present species comes boldly to the decoys if it intends to approach them, and often is so intent upon its wooden counterfeits that it has no eyes for anything else, and will fly right in, though possibly the sportsman may be standing motionless in the blind. But no Duck can get on the wing and be in full flight quicker than a Canvas Back, and many has been the disappointed gunner who, vainly imagining he was sure of his shot, but was taking time to be certain of his aim, has seen both charges from his gun strike the water behind the bird, whose mighty spring and rapid action had already carried it much farther and more quickly than its would-be captor had imagined. None can aim at a passing Canvas Back with any chance of stopping it in full flight. If there ever was a time when to " hold well ahead " was imperative, it is when shooting at this Duck passing by, or quartering.

Although this species comes so boldly to decoys, there are other times when nothing will induce it to draw near them, and then all the best imitation of its note and the frantic efforts of the concealed sportsman to attract its attention are unavailing. It simply goes upon its way, utterly indifferent apparently to the society of its fellows. Occasionally an individual will swing toward the decoys without stopping his speed for a moment, as if telling them that he knew they were there, and that they had better follow him, but giving not the slightest indication of any intention to halt. It is such times as these that try the sportsman's nerve and skill, and to stop by a well-aimed shot, and roll over one such bird stone dead in the air, when whirling along at such terrific speed, compensates him for a number of previous misses.

The Canvas Back is a brave bird, and fears no enemy of the air, possibly depending in a measure upon its great skill in diving. If a Bald Eagle comes sailing over a raft of Ducks floating on the broad water, as I have often witnessed, the birds will rise in one vast cloud and go circling about, settling after their dread enemy has passed on. But the Canvas Back is not in the cloud, nor do flocks of that bird swell its dimensions, but it keeps quietly about its occupations in company with the Swan and Geese if any are present, utterly indifferent to the movements of the other Ducks. The call of the Canvas Back is the same harsh guttural note as that uttered by the Red Head, and is usually heard when the birds are gathered together on the water. When flying it is generally silent, although sometimes it will utter this note when approaching decoys or other Ducks rafted on the sounds or rivers. This species does not bear many popular names in addition to that at the head of this article. Occasionally it is called " Canvas," simply, or White Back and Bull-Neck, and in the vicinity of New Orleans, Canard Cheval or Horse Duck.

Although, as I have already said, when this Duck has fed for a time on the wild celery its flesh is superior to that of all other Fowl, yet, when deprived of this, it is about as poor a bird as flies, not equal in any way to the Mallard or other mud Ducks that obtain their usual food where they may. It is this fact that makes such a difference in Canvas Backs when served on the table. Only those brought from localities where the wild celery grows have any qualities superior to the ordinary run of Ducks. It is generally supposed that only Canvas Backs from the Chesapeake are exceptionally fine, and they must be brought from those far-famed flats, for their delicacy and flavor to be known and appreciated. But no greater mistake can be made, as there are many places, especially among the lakes in the West, where the wild celery grows in profusion, and the Canvas Backs from those localities are equal, in gastronomic qualities, to any fed and killed on the Chesapeake.

It has seemed to me that this species has become much scarcer in the past few years; certainly many places where it used to be abundant in the winter are now almost deserted by this Duck; but it cannot be wondered at if it is so, for when we consider the persecutions it suffers from gunners striving to obtain the high price it brings in market, and the thousands that are shipped to Europe,—poor things that have been kept frozen or packed in ice until all the flavor has departed,—it is surprising that there are many left. With no effort made to preserve it from extinction, but every kind of scheme employed for its destruction, we must become accustomed to witness the noblest Game Duck that ever flew gradually disappear from our land.

ARISTONETTA VALISNERIA.

Geographical Distribution.—North America generally. Breeding from northwestern States northward.

Adult Male.—Top of head and feathers at base of bill and chin, black; rest of head and neck, brownish red. Upper part of back, chest, rump, upper and lower tail coverts, black. Rest of plumage, white, vermiculated on back, and anal region, with black. Wings similar to those of the Red Head. Bill, sloping gradually from outline of head, widening very slightly toward the end and longer than head, black. Tail, black with a grayish luster. Iris, red. Legs and feet, plumbeous. Total length, about 20 inches; wing, $9\frac{1}{10}$; culmen, $2\frac{4}{10}$; tarsus, $1\frac{7}{10}$.

Adult Female.—Head, neck, chest, and upper part of back, umber brown, darkest on top of head. Rest of back, scapulars, and sides, dark brown; tips of feathers vermiculated with ashy white. Rump, seal brown. Upper tail coverts vermiculated with yellowish brown. Tail, dark brown on outer, ashy on inner webs. Greater wing coverts, slate; outer webs of secondaries, bluish gray. Bill, black. Legs and feet, plumbeous. Under parts, white or yellowish white. Total length, 20 inches; wing, 9; culmen, $2\frac{1}{4}$; tarsus, $1\frac{1}{4}$.

RED HEAD.

THIS well-known and highly esteemed bird was at one time very abundant in many parts of North America, but constant persecution and indiscriminate slaughter of both adult and young have greatly reduced its numbers throughout the land, and in many localities where, in former times, it was very abundant in winter, it no longer appears. It is a companion of its famous relative the Canvas Back, and frequents similar localities, and seeks the same food. The distribution of the Red Head is general throughout North America, but it is not so plentiful on the Pacific side of the continent as it is in many places on the eastern coast. It does not seem to penetrate into Alaska, but it breeds throughout the so-called " Fur Countries," east of the Rocky Mountains and north of the fiftieth parallel. It also breeds in various parts of the United States along the Canadian border, but on account of the advent of railroads and increasing settlement of the country, the breeding grounds of many species of Ducks within our borders have become much restricted, and many localities formerly resorted to by the birds during the nesting season have been abandoned entirely. Absolute freedom from intrusion by depredators and security from persecution are the main requisites demanded by Water Fowl for their breeding grounds, and when these are no longer obtainable the locality ceases to be available for the purpose.

The Red Head breeds in what may be termed colonies, with many nests placed close together. These are always near the water, slightly elevated, and composed of

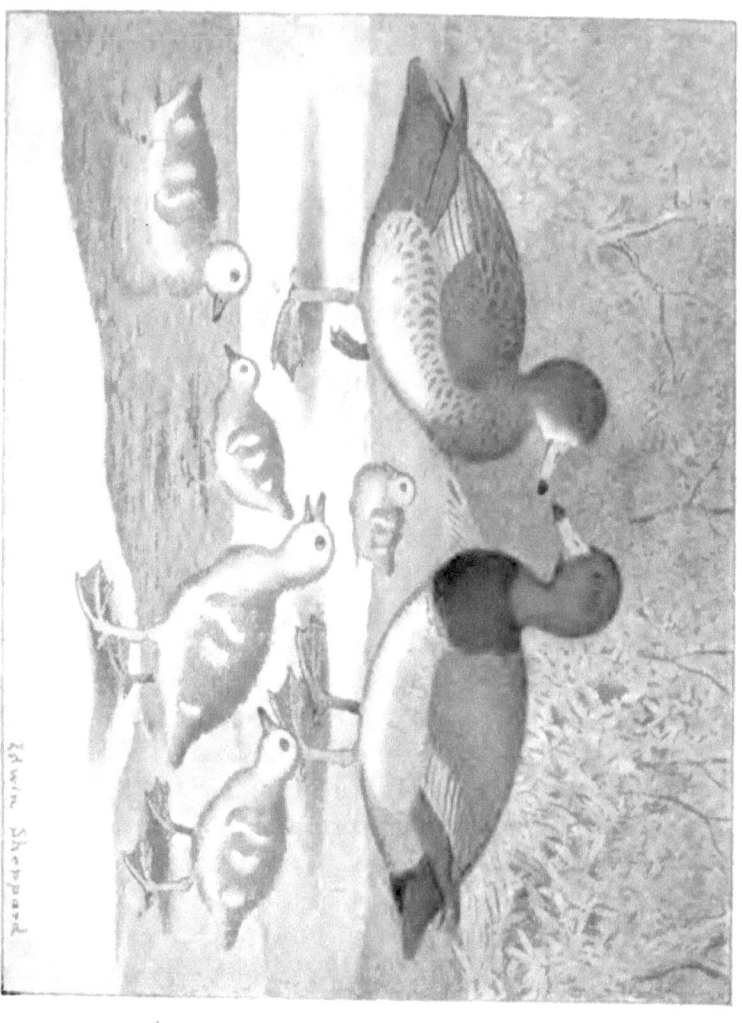

37. Red Head.

grass and weeds placed loosely together. The eggs are a creamy grayish white, and usually ten or a dozen make the full complement. This Duck has been found breeding near Calais, Maine, and also on Lake Horicon, Wisconsin, and it is thought that at one time it nested in the Sacramento Valley.

The Red Head makes its appearance, arriving from its northern resorts, where it has passed the summer, in October in large flocks. The birds fly high, in a wide V-shaped line, and proceed with great speed, accompanied by a whistling swish of the wings, so that one, even at a considerable distance, can clearly

> " Hear the beat
> Of their pinions fleet,
> As from the land of snow and sleet
> They seek a southern lea."

The flocks rarely alight at first, even when there may be numbers of Duck congregated on the water, but traverse the length of the sound or lake as if reconnoitering the entire expanse, and trying to select the best feeding ground. After having passed and repassed over the route a few times, the flock begins to lower, and gradually descending, at length the wings are set and the birds sail gradually up to the chosen spot, usually where other Duck are feeding, and drop in their midst with many splashings. But while this is the usual method adopted by newcomers, sometimes the programme is changed and the birds, attracted by a large concourse of their relatives, particularly if the day be calm and the sun shining with considerable heat, will suddenly drop from out the sky in a rapid zigzag course, as if one wing of each Duck had been broken, and they cross and recross each other in the rapid descent, their fall accompanied by a loud whirring sound, as the air is forced between the primaries. On such occasions the flock is mixed all up together in a most bewildering manner, until, arriving a few feet above the water, the wings become motionless and the birds glide up to and alight by the side of their desired companions.

Early in the morning, and again late in the afternoon, the Red Head regularly takes a "constitutional." The flocks, that have been massed together during the night or the middle of the day, rise from the water, not all together but in companies of several dozen, and stringing themselves out in long, irregular lines, each bird a little behind and to one side of its leader, fly rapidly up and down, at a considerable height over the water. Sometimes these morning and evening promenades are performed at a great elevation, so that the movement of the wings is hardly perceptible. On such occasions they appear like a dark ribbon against the sky, and the comparison is strengthened by the fact that every movement of the leader elevating or depressing his course is imitated exactly by all those which follow, and so the line has frequent wavy motions like currents passing through it, as when a ribbon is held in the fingers and a flip given to it which causes it to undulate along its whole length.

This species is a deep-water Duck and keeps out in the center of rivers or lakes, congregating at times in such numbers as to form immense rafts; hence it is sometimes called "Raft Duck." It dives readily and to considerable depths, and pulls up the grass and roots found on the bottom, returning to the surface to enjoy the fruits of its labor, and not infrequently to find them snatched away by the ever-active Widgeon, always on the lookout for tid-bits it is unable to dig up for itself. Great flocks of these birds are always in attendance on the Red

Heads and Canvas Backs, and secure a large proportion of the food these diving Ducks send to the surface. Red Heads feed much at night, especially if the moon is shining, and at such times are exceedingly busy, and the splashing of diving birds, the coming and going of others, and the incessant utterings of their hoarse note, are heard from dark to daylight. They also feed by day, if the weather has been stormy, but on quiet, pleasant days they rarely move about much, but remain quietly out in the open water, sleeping, or dressing their feathers, or occasionally taking a turn beneath the surface as though more in an exploring mood, than for the purpose of seeking food. In localities where the marshes are scattered throughout the broad sounds, or form the banks of the rivers, the Red Heads are accustomed to resort to them a great deal, paddling close to the edges looking for insects or other animated objects suitable for food, or frequenting the ponds, when such exist, in company with mud Ducks and others which habitually seek such places.

As a rule the Red Head is gentle and unsuspicious, and readily comes to decoys. It has a habit on such occasions that causes great destruction to the flocks. When the birds have sailed up to the blind and either are preparing to alight, or hesitating whether or not to go on their way, the members crowd close together, or "bunch," as it is called, giving the sportsman an opportunity to discharge the contents of his gun into their midst with the effect of killing a number of birds and wounding many more. An injured Red Head is not an easy bird to capture, as it dives and skulks with great rapidity and skill, and if on open water always moves against the wind. If near a marsh, it will get under the bank, or crawl up into the grass, and it needs a good dog to find it. When all other means fail it will dive to the bottom, seize some grass in its bill, and hold on until life is extinct; commit suicide by drowning, in fact, rather than fall into the hands of its pursuer.

Sometimes this duck is known as Gray-Back, and in Louisiana as Dos Gris, the French equivalent for the same name, and also Canard Violon. The Red Head bears confinement well, but does not breed readily when domesticated. The note of this species is a hoarse guttural rolling sound, as if the letter R was uttered in the throat with a vibration of the tongue at the same time. It is easily imitated, and the bird readily responds to the call of its supposed relative. Some other ducks, like the Canvas Back, different species of Scaup Ducks, Sprigtail, etc., have a similar call. The flesh of the Red Head, when it has been feeding upon wild celery and such dainty food, for tenderness and flavor is excelled by no other Duck, and many are passed off for Canvas Back. I have tried both, shot the same day on the Chesapeake, and the birds had doubtless fed on the wild celery, and in point of excellence there was no difference between them. Occasionally I have found the Red Head the better bird of the two, but this was exceptional. Of course, if the heads are served with the body, there is never any difficulty in distinguishing them, provided the heads really belong to the bodies served, but in all cases the Canvas Back is considerably the larger Duck. A knowledge of comparative anatomy is very useful in a case where a decision as to the identity of these Ducks is required, as the result may mean a difference of quite a sum of money to the host, for probably more so-called Canvas Backs and even Red Heads are eaten during one winter in our country than ever flew within its limits at the same period. This species has various names, some of which are American Pochard, Raft Duck, and Red-headed Raft Duck.

ÆTHYIA AMERICANA.

Geographical Distribution.—North America generally. Breeds from California and Northern tier of States to the Arctic regions.

Adult Male.—Head, full and puffed out, and with the neck is rich reddish chestnut, glossed at times with reddish purple. Lower neck, chest, upper parts of back, rump, and upper and lower tail coverts, black. Back, scapulars, sides and flanks, grayish white, finely undulated with black. Wing coverts, ash gray. Speculum, ash gray, bordered above with black and posteriorly with white. Primaries, dark brown on tips and outer web, gray on inner. Tail, dark brown. Under parts, white, growing darker toward the under tail coverts. In some specimens the under surface is whitish brown. Bill, broad, flattened, widest at tip, rising at base abruptly to the forehead, forming a very different angle to the bill than that of the Canvas Back, dull blue in color, and crossed by a black bar near the tip. Iris, orange. Legs and feet, grayish blue; webs, dusky. Total length, about $19\frac{1}{4}$ inches; wing, 9; culmen, $1\frac{9}{10}$; tarsus, $1\frac{5}{10}$; bill at widest point, $\frac{8}{10}$.

Adult Female.—Head and neck, pale brown; darkest on top of head. Chin and throat almost white, as is also, in some specimens, the loral space. Cheeks, frequently grayish brown. Back and scapulars, grayish brown; feathers, tipped with light gray; wing coverts and secondaries, pearly gray; speculum, light ash gray. Secondaries, pearly gray on outer webs; edged with black. Primaries, fuscous on outer webs; dark buff along the shafts and on inner webs. Lower back, blackish brown, lighter on upper tail coverts; feathers of latter, tipped with pale brown. Chest, sides, and flanks, grayish brown; feathers, tipped with fulvous. Bill, pure white; anal region and under tail coverts, brownish white, darker on the thighs. Bill, pale blue, black at tip. Legs and feet, grayish blue. Total length, 19 inches; wing, 9; culmen, $1\frac{7}{8}$; tarsus, $1\frac{3}{4}$.

Downy Young.—Sides of head and neck, and lower parts, deep buff, palest on the belly. Top of head and upper parts of body, ochraceous olive brown, with a yellow spot on sides of body and rump, and on border of wings.

SCAUP DUCK.

THE various published accounts of this species fail to give a complete history of its economy and habits because this Duck and the Little Scaup, which so much resembles it, have been by nearly all authors greatly confused together. So far as my experience enables me to judge, the Big Black Head is a bird that mostly frequents the coasts, and is not so often found inland as its relative, which at times is very abundant on our lakes and rivers, and those writers who have mentioned this bird as being so very common in many localities in the interior of the United States probably really have reference to the Little Black Head, quite another species.

This Duck breeds in the far North, from Alaska on both sides of the mountains across the continent, and possibly to the vicinity of the Arctic Sea. It is also an inhabitant of the Eastern Hemisphere, and is found from the British Islands to China breeding in the northern portions, but not south of the latitude of Lapland. It is found on Kotzebue Sound, Alaska, and on the Yukon River it is plentiful in summer, and is also dispersed throughout its Delta, and along the islands of the Aleutian chain. The birds arrive at their breeding grounds from the South early in May, and scatter over the marshes and numerous small ponds, and select their mates preparatory for the nesting season rapidly approaching. A place amid the high grass, close to the water, is selected for the nest, so close indeed that the bird can swim to it. Loose grass, lying about, is

38. Scaup Duck.

gathered together, and down, plucked from the bird's own breast to form a bed, is placed upon it, and from six to eight eggs, rather a small complement for a Duck, are deposited. These are pale olive gray in hue, and are hidden in the downy covering whenever the female is off the nest. June is the month for incubation, and the period of hatching must be from three to four weeks, for in August half-grown young are seen. As soon as the ducklings escape from the egg, they are led by the mother to some large body of water, where frequently several broods unite and form quite a flock.

In October the Big Blue Bill enters the limits of the United States, coming from its Northern home, and is found along the coast of both oceans, going as far south as Mexico during the winter. It flies with great swiftness, and is a most expert diver; a wounded bird, unless very badly crippled, being practically impossible to capture. The Bay Broad Bill, as it is sometimes called, does not go in such large flocks as is the habit of its smaller relative, and keeps a good deal about the coves and marshes. It decoys readily, and utters at times a note similar to the guttural sound made by the Canvas Back, Red Head, and other diving Ducks. I have not noticed that it associates much with the Little Broad Bill, but keeps to the society of its own species, and goes in flocks usually of less than a dozen members. At no time have I ever seen it rafted in the open water in great numbers, as frequently is the case with the Little Black Head.

The present species is quite a large Duck, and has a metallic green luster on the feathers of its head and neck, which enables it easily to be discriminated from the purple-hued head of the allied form. It bears many popular names among the gunners throughout the United States, a number of which are the same as those of its small relative, with a prefix denoting bigness; thus, in addition to those already given, it is called Big Black Head, Big Scaup, Big Shuffler, Big Broad Bill; also Salt-Water Broad Bill (indicating its preference for the seacoasts), Bay Broad Bill, Gray Back, Black Neck, Dos Gris in Louisiana, and various others, some of which are purely local, and rarely heard. As a bird for the table it is about on a par with the Little Scaup Duck, and, when it has fed upon wild celery and other tender grasses, its flesh is well-flavored, but if away from localities where these grasses are found, it is not very particular upon the quality of its diet, and often has a fishy and rank flavor, not in any way desirable. I do not regard it as common a species as many of the diving Ducks found within our borders, and the days when I have met with them, even in comparatively large numbers, have been exceptional. For a long period its distinctness from the Small Broad Bill was unknown, and for some time after it was suggested that there were two species, both ornithologists and sportsmen were skeptical of the fact.

FULIGULA MARILA.

Geographical Distribution.—North America generally, south to Guatemala. Also in Northern portion of Old World to China. Breeds in Alaska, and in the Arctic regions east of the mountains.

Adult Male.—Head, neck, fore parts of back and chest, black, with green reflections on head and neck. Lower back, rump, upper and under tail coverts, also black. Middle of back, scapulars, sides, flanks, and anal region, white, undulated with fine black lines. Wing coverts, blackish, finely barred with white. Speculum, white, bounded in front by a black line formed by the tips of the greater coverts. Tertials, black, glossed with green; some of the large ones vermiculated with white. Primaries, dark brown, with black tips, and a grayish or whitish area on inner webs. Tail, blackish brown. Belly, white. Bill, bluish gray; nail, black. Iris, yellow. Legs and feet, plumbeous. Total length, about 19 inches; wing, $8\frac{1}{10}$; culmen, 2; tarsus, $1\frac{4}{10}$.

Adult Female.—Forehead, and sides of bill at base, white. Rest of head, neck, and breast, snuff brown. Upper parts, dusky brown; tip of feathers, lighter. Back and scapulars, vermiculated slightly with white. Wings, purplish brown, with white speculum. Primaries, with the tips and outer webs of first two, blackish brown, remainder, pinkish buff, or whitish brown, the latter showing like a patch when wing is closed. Flanks, brown, vermiculated with white. Belly, white. Anal regions and under tail coverts, dark brown, inclined to an olive shade; feathers, tipped with white. Tail, dusky brown, lighter than the rump. Iris, bill, legs, and feet, colored as in the male. Total length, about 19 inches; wing, $8\frac{1}{10}$; culmen, $1\frac{3}{4}$; tarsus, $1\frac{4}{10}$. There is very little if any difference in the average size of the sexes of this species.

LESSER SCAUP DUCK.

LITTLE Broad Bill, Little Black Head, Little Blue Bill, Shuffler, River Broad Bill, Black Head, Creek Black Head, Broad Bill, Raft Duck, and Flocking Fowl are some of the names by which this species is known in various parts of our country. It is one of the most common of our Ducks, and it appears to me to be growing more abundant; at all events, this is so in many localities. Whether this is caused by an actual increase in numbers, or that the birds have merely frequented localities usually neglected by them, and so seem to be more numerous, I cannot say. The species has a wide distribution, ranging over the whole of North America, and going south in winter as far as Guatemala and the West Indies. It breeds north of the United States, mainly in the Arctic regions and also, possibly, in Minnesota, and perhaps in some other of the border States; but whether it goes west of the mountains in the Territory of Alaska is difficult to determine, as by many writers this bird and the previous species have been so generally regarded as the same, that it is impossible to decide by their narratives which one is intended. Dall and Kennicott say it breeds plentifully on the Yukon River, while Nelson, an equally competent observer, states that during a long residence, at the Yukon mouth and to the northward, he did not see a single individual of the Little Scaup, although the Big Scaup was abundant, and Turner does not mention it among the birds seen by him in Alaska. From this

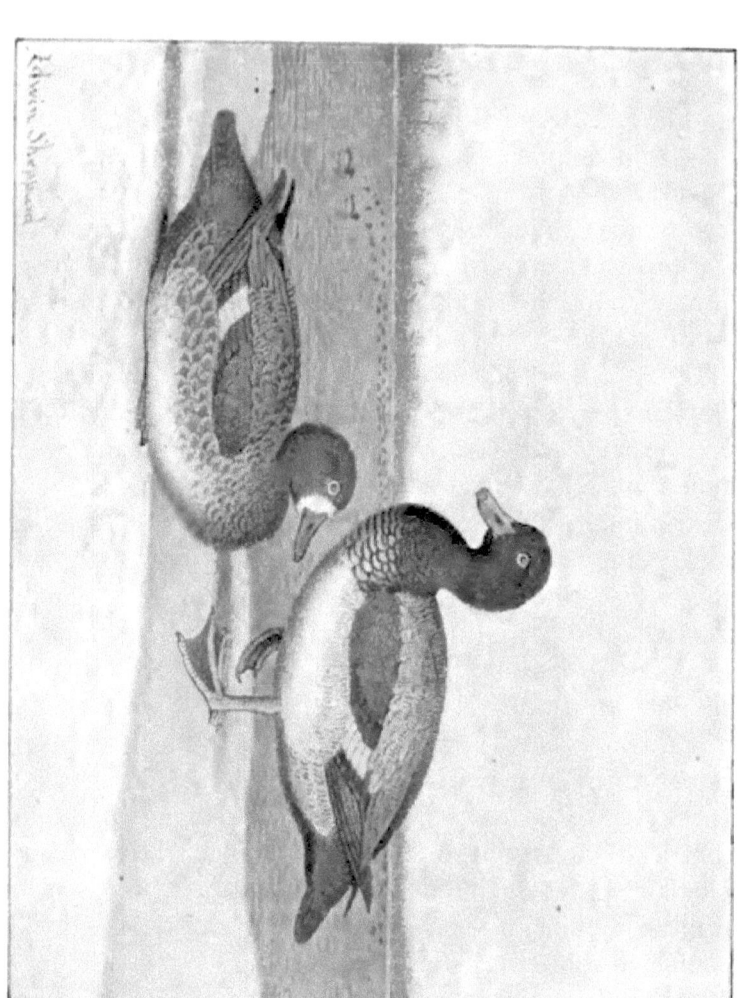

39. Lesser Scaup Duck.

it would be fair to infer that this species breeds on the eastern portion of the Arctic region, and if present at all, is an exceptional visitor within the limits of Alaska.

The nest, found on the lower Anderson River by MacFarlane, was placed in a swampy tract near a wooded country, and was simply a depression in the center of a tuft of grass, and lined with down, probably from the female's breast. Another was placed in a clump of willows in the midst of a swamp, and close to a small lake. The location of these nests were somewhat different from those chosen by the Big Scaup Duck, which, as already stated, were almost in the water, or so near that the female could swim to and from it. The eggs, usually nine in number, are a pale grayish buff sometimes tinged with olive. The male keeps in the vicinity of the nest, but it is not known that he shares in any of the duties of incubation.

The Little Broad Bill is a cold-weather Duck, and is frequently observed flying about when the ponds and rivers are nearly all frozen over. At such times it visits the air holes, and is very busy diving for food, which it brings up from the bottom. It arrives within our borders rather late in the autumn, and keeps in large flocks in the center of the broad water, away from the shore. It is one of the most expert divers among the Duck tribe, and can reach the bottom to pull up grasses or pick up mollusks, in as deep water as any of its relatives, no matter how skillful they may be in the business. Like the Canvas Back and other species which frequent deep water the Little Black Head propels itself beneath the surface by its wings, using the webbed feet merely as rudders.

This Duck is very tenacious of life, and it requires a hard blow, and shot of considerable size, to kill it.

When wounded it shows much cunning, skulking and hiding among the grass, or beneath the overhanging banks of marshes, and it will immerse its entire body beneath the surface, leaving only the bill exposed and, if all else fails, will go to the bottom and hold on to the grass until life is extinct. The Little Broad Bill is very swift upon the wing, and comes to decoys readily, but can get away from their vicinity when alarmed about as quickly as anything that flies. It generally goes in flocks of from one to three dozen, sometimes considerably more, and comes boldly up to a blind or sink-boat, usually " company front," and on the discharge of a gun the birds scatter in every direction like a swarm of bees, straight up in the air, or off to either side in most admirable confusion, gathering together again when the point of danger is passed, and speeding onward in undulating lines over the middle of the broadest stretch of water. The wounded birds that have fallen amid the decoys immediately dive, sometimes going directly under water from their descent in mid-air, appearing again only for a second at some distance away, either headed for the nearest marsh, or swimming in the wind's eye toward the open water. When wounded they are very difficult to capture and bother even the best retriever greatly; diving incessantly and with great rapidity, sometimes right under his nose, appearing behind him or on one side, and keeping the dog spinning around like a top in his efforts to sieze such a slippery object.

The flesh of this duck is sometimes very tender and of good flavor, but these qualities depend altogether upon what it has been feeding, for it is not very select in its diet, and will swallow all kinds of shell-fish, probably frogs, newts, or any similar creature it can catch, and on this food it becomes rank and disagreeable, quite unfit for the table. But if fed upon roots of water plants, wild rice, celery, or other similar tender grasses, it is a very good little bird indeed. For the sportsman there is no better object upon which to try his skill than this Duck; its rapid flight and quick, unexpected movements on the wing frequently bringing to nought the efforts of the most expert gunner.

Considerable variation among individuals of this species exists in their measurements, and occasionally they approach in size those of the Big Scaup, so that, as regards the females, it is at times very difficult to distinguish which species is represented. Adult males can easily be identified, no matter what their dimensions may be, the metallic hues of the head making them readily recognizable. But there is little in the coloring of the females to separate them from the larger species, and if the wing should exceed eight and one-quarter inches in length it is exceedingly difficult to say to which form the bird should be referred. The company the specimen kept when it was killed, if that could be ascertained, would be the surest test for identification, as these two Scaups are rarely found associating together. The eggs also vary greatly in their measurements.

FULIGULA AFFINIS.

Geographical Distribution.—North America generally. Breeding north of United States. In winter to Guatemala and the West Indies.

Adult Male.—Head, neck, and fore part of body, black, with purple reflections on head. Back and scapulars, white, barred with narrow irregular black lines. Wing coverts, dusky, mottled with white. Secondaries, white, the tips, black, with a greenish gloss, forming a white patch or speculum on the wing. Tertials, black, glossed with green. Primaries, brown, blackish at tips and toward edges of the webs. Rump and upper tail coverts, black. Breast and abdomen, white. Flanks, white, barred with irregular black lines, more or less distinct. Crissum, dusky, mottled with white; under tail coverts, black. Tail, black. Bill, bluish white, nail, black. Legs and feet, light slate or plumbeous. Iris, yellow. Length, about 16 inches; wing, 8; tail, 3, tarsus, $1\frac{8}{16}$; culmen, $1\frac{7}{16}$.

Adult Female.—Space about base of bill, white. Rest of head and neck, snuff brown. Upper back and breast, amber brown; the feathers, margined with pale brown on the former, ochraceous on the latter. Back and scapulars, fuscous, mottled with white. Wings, dark brown; speculum, white. Flanks, dark grayish brown, tips of feathers, white. Under parts, white. Rump and upper tail coverts, dark grayish brown. Anal region and under tail coverts, pale grayish brown, much lighter than rump and upper tail coverts, and grading into the white of the abdomen. Tail, dark grayish brown, edges of webs, ochraceous. Bill, legs, and feet, colored as in the male. Size similar to that of the male.

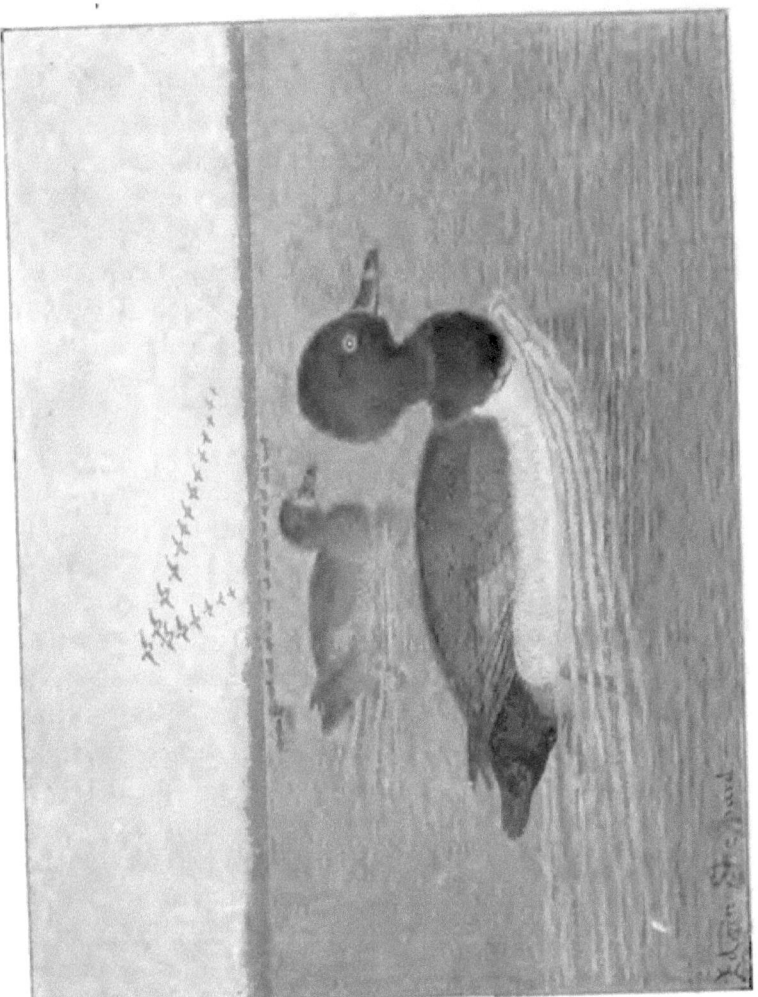

40. Ringed-Neck Duck.

RINGED-NECK DUCK.

NOWHERE so abundant as is the last species, the Ringed Neck has nevertheless as wide a distribution as the Broad Bill, and ranges over the whole of North America from the Arctic Sea to Guatemala and the West Indies. It bears a considerable number of names, and is often confounded with the Little Scaup Duck, and in different portions of the United States is called Tufted Duck, Ring Bill, Bastard Broad Bill, Shuffler, Ring-billed Shuffler, Ring-billed Black Head, Ringed-neck Black Head, Ringed-neck Scaup, and Canard Noir in Louisiana. It probably has some more names, but these are about all that I have heard applied to this bird myself, and those most commonly used are the one at the head of this article, and Ring Bill.

It is not a common species, goes in small flocks, and frequently is found in company with the Little Scaup, which it resembles very closely in its general habits. It breeds from the northern part of the United States northward, and has been seen in Alaska, but the nest has not yet been found there, although it is very probable that the species does breed in that Territory. The few examples seen were so shy that it was impossible to get near them. Nests of this Duck have been found in Wisconsin and Minnesota. In the former State one was found on a bog in thick cover near Pewaukee Lake, and was formed of grasses, and lined with feathers. The nesting habits of this Duck are not very well known and its breeding limits have not been ascertained. The eggs are grayish white, sometimes with a buff tinge. My friend Mr. George A. Boardman found a nest of this species, containing eleven eggs, on the St. Croix River near Calais, Maine, and on another occasion secured a brood of ducklings together with the old ones. This would seem to show that the Ring Neck probably breeds along the northern border of the United States wherever suitable localities occur.

This Duck is more often seen on the rivers and inland lakes than on the sea-coast, although it is found every winter near the borders of both the Atlantic and Pacific oceans. Its flight resembles that of the Little Blue Bill and is quite as swiftly performed, and its movements on the wing are equally as quick as those of its relative. It comes readily to decoys and is as tenacious of life and as skillful in evading pursuit, when wounded, as is the Little Scaup.

The Ring Neck resembles the Little Broad Bill in general appearance, but is a much handsomer bird, the peculiar white marking upon the bill, and the red ring, more or less distinct, around the neck, making it very conspicuous. As a bird for the table it is about equal to the Little Black Head, and what has been already said in this respect of that species is equally applicable to this one. I think it is more plentiful on the waters of our Western States, especially those in the Valley of the Mississippi, than it is anywhere in the East. On the Pacific coast it goes from Mexico to northern Alaska, but is nowhere very abundant. Like the Little Scaup this is a cold-weather Duck, and unless everything is entirely frozen up, occasionally remains in northern latitudes all winter.

FULIGULA COLLARIS.

Geographical Distribution.—North America from the Arctic Ocean to Guatemala and the West Indies. Breeds from northern United States to limit of its range in Arctic America.

Adult Male.—Head, neck, breast, upper parts, and under tail coverts, black, with a gloss of purple on head, and greenish on back. A more or less distinct chestnut collar around middle of neck. A triangular white spot on chin. Wings, blackish brown, with a green gloss. Speculum, gray. Under parts, white; the flanks and sides, waved with fine black lines. Crissum, dusky, mottled with white. Bill, black, with the base, edges, and a bar across maxilla near nail, pale bluish. Legs and feet, grayish blue, webs, dusky. Iris, yellow. Length, about $17\frac{1}{4}$ inches; wing, 8; tail, $3\frac{4}{10}$; culmen, $1\frac{9}{10}$; tarsus, $1\frac{1}{4}$.

Adult Female.—Top of head and back of neck, dark brown; sides of head, grayish white. Loral space, forehead, eyelids, chin, throat, and neck in front, yellowish white. Sides of neck, light brown. Back and wings, dark brown, feathers margined with rufous. Speculum, gray; outer webs of outer tertials, metallic green. Lower back and rump, black; upper tail coverts and tail, pale brown, feathers, tipped with yellowish brown. Upper breast, sides, and flanks, fulvous brown, tips of feathers, yellowish brown. Lower breast and belly, white. Anal region, dull brown; under tail coverts, white, speckled with brown. Bill, slate, with pale blue band crossing it near tip. Total length, 17 inches; wing, $7\frac{1}{2}$; culmen, $1\frac{3}{4}$; tarsus, $1\frac{1}{4}$.

Downy Young.—Top of head and neck behind, dark grayish brown; ears, grayish brown; rest of head and neck and lower parts, pale buff; upper parts, grayish brown, with a spot in center of back and on each side of back and rump, and a bar across posterior border of wings, light buff.

LABRADOR DUCK.

FORMERLY not an uncommon bird along the Atlantic coast as far south as Delaware, the Labrador Duck has, for over twenty years, ceased to make its appearance anywhere within our boundaries, and it would seem that, from some reason quite inexplicable, it has become extinct. The Pied Duck, as it was sometimes called, fifty years ago was said to be frequently offered for sale in the markets, hanging among strings of other species of Ducks. It was not known to Wilson, and Audubon never saw it alive; the birds from which he made his drawing having been killed by Daniel Webster on Vineyard Island, coast of Massachusetts. This pair is now in the collection of the National Museum at Washington. Very little is really known about the habits of this species. There are no authentic accounts of its nest or eggs, and it is doubtful if anyone, save perhaps an Eskimo, has ever seen either one or the other. John W. Audubon had several deserted nests shown him at Blanc Sablon, Labrador, as belonging to this Duck, but he saw no individuals, and it may be seriously doubted if the Labrador Duck ever had anything to do with them.

Fifty years ago, according to Giraud, this bird, known to the gunners of Long Island as the Skunk Duck, on account of its peculiar black and white markings, was even then very rare. The people of the New Jersey coast called it "Sand-shoal Duck." It was said to feed on shell-fish, which it procured by diving. Between 1860 and 1870 I saw at various times a con-

41. Labrador Duck.

siderable number in Fulton and Washington Markets of New York. They were mostly females and young males, a full-plumaged male being exceedingly rare. Sometimes there would be as many as a dozen hanging together, and then weeks might elapse before any more were seen. At that time, while it was remarked that it was a curious circumstance that only females or young males were to be had, no one imagined that the species was approaching extinction; for when immature birds existed there must be both parents somewhere. Gradually, however, the specimens became fewer, and appeared at longer intervals, until they disappeared entirely. During the twenty years between 1850 and 1870 a few full-plumaged males were obtained, and one of the finest I ever saw I bought from a taxidermist in Brooklyn, who had it at the time in the flesh. During the periods of which I speak, there would have been no difficulty in procuring quite a large series of females and young males, but as it was supposed these could be obtained whenever wanted, they were neglected.

The cause of the disappearance of this Duck no one knows. Various attempts have been made to account for it, but none has been satisfactory. By some naturalists it is conjectured that it was brought about by the destruction of the eggs, but we have no reason to suppose that any more eggs of this species were destroyed, from any cause whatever, than were those of any other Duck. It was not exterminated by man with the gun, for he did not get a chance—the birds gave him too few opportunities. Being strong of flight as well as a skillfull diver, there was no reason why, if necessary, it could not have easily and rapidly conveyed itself away from any threatened danger, and no matter how the fact of its extinction is regarded and what were its possible causes, no explanation can be given that is satisfactory. It is one of those inexplicable phenomena that occasionally arise to perplex and baffle the best informed person. As a bird for the table, as might have been expected from its choice of food, it was not very desirable, being fishy and of a strong flavor; evidently only on a par with the usual run of Sea-Ducks. About forty specimens only are known to be preserved in all North America, and not half that number in all Europe. The finest collection of these birds in the world is in the New York Museum of Natural History, where seven adult males, females, and young males are to be seen. Five have been artistically mounted in a group with characteristic surroundings of ice and water (for it was a cold-weather bird), forming one of the rarest and most valuable ornaments in the possession of the museum. While we marvel at the disappearance of this bird from our fauna, similar or equally forcible methods are at work, which in the process of time, and short time too, will cause many another species of our Water Fowl to vanish from our lakes and rivers, and along the coasts of our continent. Robbing the nests for all manner of purposes, from that of making the eggs an article of commerce or posing as specimens in cabinets, slaying the ducklings before they are able to fly and have no means of escape from the butchers, together with the never ceasing slaughter from the moment the young are able to take wing and start on their migration, at all times, in all seasons, and in every place, until the few remaining have returned to their summer home, all combined, are yearly reducing their ranks with a fearful rapidity, and speedily hastening the time when, so far as our Water Fowl are concerned, the places that now know them, and echo with their pleasant voices, shall know them no more forever.

CAMPTOLÆMUS LABRADORIUS.

Geographical Distribution.—Formerly on the Atlantic coast from New Jersey northward. Now extinct.

Adult Male.—Head, neck, breast, scapular, and wings, except primaries, white. Stripe on crown and nape, ring around lower neck, back, rump, primaries, upper tail coverts, tail, and entire lower parts, black; the tail with a grayish tinge. Cheeks, frequently yellowish white. Long scapulars, pearl gray; tertials, with black edges. Bill, black, blue along the base of culmen, and orange at base and along edges of maxilla and mandible. Iris, reddish brown; feet and legs, grayish blue. Total length, about $29\frac{8}{10}$ inches; wing, $8\frac{7}{10}$; culmen, $1\frac{2}{4}$; tarsus, $1\frac{1}{2}$.

Adult Female.—General plumage, uniform brownish gray. Tertials, silvery gray, edged with black. Secondaries, white, forming a speculum, inner secondaries with black edgings. Total length, about 18 inches; wing, $8\frac{4}{10}$; culmen, $1\frac{6}{10}$; tarsus, $1\frac{4}{16}$.

Young Male.—Very similar to adult female, but the chin and throat, pure white, and in some specimens the breast also, but in others the white of this part is merely indicated. The greater wing coverts are also sometimes white.

GOLDEN EYE.

CIRCUMPOLAR in its distribution, and ranging throughout the whole of North America from the Arctic Sea to the island of Cuba in the south, and from the Atlantic coast across the continent to the Pacific, the Golden Eye is among the best known of our Ducks. It breeds from about the parallel of Massachusetts northward to the Arctic circle in the interior, and is rare upon the coast, though in some of the Aleutian Islands a few remain all winter. It is a hardy bird and able to withstand severe cold. The Golden Eye breeds in the hollows of trees, the entrance often appearing to be absurdly small for the size of the Duck, but like other web-footed tree-nesting species it finds no difficulty in entering its chosen abode. The eggs are a pale grayish green color, and from six to eight seem to be the full complement. This beautiful Duck is known to many as the "Whistler," and beside this name it is called Spirit Duck, Whistle Wing, Whiffler, Great Head, Bull Head, and Plongeur in Louisiana. Its principal appellation, of Whistler, is given on account of the shrill noise the wings make when the bird is flying; a sound so sharp and penetrating that the species is indicated long before it comes clearly into view.

The Golden Eye rises directly from the water, but not with a spring like the Mallard or Dusky Duck, flying low at first, but rapidly rising until it has attained a lofty altitude, when it moves on in a straight line, and, from the first motion made to leave the water, the loud

42. Golden Eye.

"whistle" of the wings is heard. The Whistler is a high flyer, and upon the sea-coast pays but little attention to decoys, although it will occasionally come to them. It is generally seen singly or in pairs, the male frequently leading the female, though at times their positions are reversed as is usually the case with Ducks when traveling in pairs, but in the interior small flocks are not uncommon, especially upon the rivers, which are much frequented by this species. The large thickly crested or rather fluffy head of this Duck is beautifully resplendent with metallic green hues, particularly noticeable when the sun's rays fall upon it, the brighter portions contrasting with those in shadow, like brilliant emeralds lying on dark green velvet. The Indians along the River Yukon stuff the skin of this Duck and ornament it with beads, and give it to a child for a doll or toy.

As a diver the Golden Eye ranks as a master. So instantaneous are its movements upon the water when disappearing below the surface, that shot from a gun cannot travel to the spot it occupied quickly enough, if the bird has seen the flash, for it is under water at once. The Indians are superstitious about it on account of its wonderful quickness, and the name of Spirit Duck was given to it by them as typifying a being endowed with supernatural powers. It is able to keep up this rapid diving for a long while, and one will waste his time if he waits hoping to catch a Golden Eye napping. This Duck feeds at the sea-coast, on shell-fish mainly, which it procures by diving, but on inland lakes and rivers it must eat grasses and roots, for its flesh has a very different flavor and is tender and delicate. In South Carolina it visits the rice-fields and feeds on the grain. It is often seen in company with the Little Broad Bill, Buffle Head, and sometimes with Mergansers, paddling along near the banks of marshes, and dabbling in the mud, sifting it between the mandibles. Not often is it in the habit of alighting in the open water away from the land, and whenever it does do so it appears uneasy, as if anticipating some unseen danger, and is one of the first Ducks to take wing should an alarm be sounded.

When the weather is stormy, heavy rains or snow, the Golden Eye keeps close to the shore, and if on a river, flies up and down near the bank. It does not seem to be so wild on stormy days, perhaps being more anxious to find a shelter from the gale, and less mindful, for the moment, of possible danger to itself from the usual causes. The Whistler is a silent bird, its wings generally providing all the noise it makes, but occasionally I have heard it utter a hoarse kind of croak similar to that made by the Merganser, but at no time anything resembling a quack. The European Golden Eye I consider specifically the same as the American bird. It has been claimed that the two are distinct, the difference consisting mainly in size, the European being somewhat smaller. It is too fine a distinction and nothing is gained by this attempt to separate the birds, scientifically or otherwise, for such an unsatisfactory reason.

CLANGULA CLANGULA.

Geographical Distribution.—North America, from the Arctic Sea to Mexico and Cuba. Breeds from Massachusetts and the British Provinces, northward. In Old World from Great Britain to Japan, and from Arctic regions to Northern Africa.

Adult Male.—A rather bunchy occipital crest, extending a short distance down the hind neck. Head and upper part of neck, glossy green, with purple reflections. A large, rather oval white spot on lower part of the lores, advancing close to base of bill. Lower part of neck, upper part of back, short scapulars, greater wing coverts, most of the secondaries and under parts generally, pure white. Rest of upper parts, long scapulars, and some secondaries, black. Base of secondaries, black, forming an indistinct bar hidden under the white tips of greater coverts. Primaries and their coverts, brownish black. Outer webs of uppermost flank feathers, partly or wholly white. Tail, ashy. Bill, greenish black. Legs and feet, orange; webs, dusky. Iris, golden yellow. Total length, about 20 inches; wing, $8\frac{7}{10}$; tail, $4\frac{1}{4}$; tarsus, $1\frac{1}{2}$; culmen, $1\frac{6}{10}$; bill, height at base, $\frac{9}{10}$; width, $\frac{9}{10}$; width of nail, $\frac{1}{4}$.

Adult Female.—Head and upper part of neck, hair brown. Collar on neck, very narrow behind; white, streaked with bluish gray. Back, blackish brown; feathers, on upper back, edged with bluish gray; those of upper tail coverts, tipped with pale brown. White on wings not so extensive as on those of the male. Tips of greater wing coverts, black, forming a bar across the white. Primaries, brownish black. A band of bluish gray across upper part of breast. Under parts, white. Thighs, dusky. Tail, dark brown, like the back. Bill, dull yellowish, shaded with blackish brown. Bills of different individuals vary in color. Legs and feet, orange; webs, dusky. Iris, golden yellow. Wing, 8 inches; tail, $4\frac{1}{2}$; culmen, $1\frac{8}{10}$; tarsus, $1\frac{4}{10}$; bill, height at base, $\frac{9}{10}$; width at base $\frac{9}{10}$; width of nail, $\frac{1}{4}$. *Height of bill from point of angle to nearest cutting edge* LESS *than the distance between the farthest edge of nostril and nearest feathers at base of bill.*

Downy Young.—Upper parts, dark brown; throat, white; breast and flanks, pale brown; belly, pale gray.

BARROW'S GOLDEN EYE.

A NEAR relative to the Common Golden Eye, the present bird, sometimes called the Rocky Mountain Garrot, is much more restricted in its range, and although it occurs in Iceland, may be regarded as essentially an American species. In the West it breeds as far south in the Rocky Mountains as Colorado, and in the East from the Gulf of St. Lawrence, northward. It has been procured at Sitka, Alaska, and noticed by Dall on the Yukon, but its appearance in that Territory is probably only exceptional in the northern portions. It breeds also in Greenland and Iceland. In winter it goes south on the Eastern coast to New York, and to Illinois, Utah, and Colorado in the West. As yet it has not been found west of the mountains south of Alaska.

For a long time this Duck was confounded with the Common Golden Eye, and supposed by some to be merely a phase of the summer dress of the well-known bird. It is an inhabitant of the interior, and I have never seen it upon any of our coasts, though it does at times visit the vicinity of the ocean. In the Rocky Mountains it has been found breeding at a high altitude and it is believed to nest in Maine. It breeds in the hollows of trees, as is the habit of the Whistler, and the number of eggs is from six to ten. They are dark grayish green in color. In Iceland, where trees are scarce, this species nests in holes in the ground, especially among the blocks and in the crevices of broken lava, in company with the

43. Barrow's Golden-Eye.

Merganser. Sometimes these holes are so deep that the eggs are entirely out of reach.

Barrow's Golden Eye is a somewhat larger and handsomer bird than the common species, with the crested head beautifully colored in metallic hues of green, blue, and violet, changing as the rays of light fall upon it. The large crescentic white mark before the eye in the male will always easily distinguish this bird from its relative, and it is to be wondered that the two were ever considered as one species. The females of the two forms are very difficult to distinguish apart, and at times will bother even an expert. The chief difference is in the bill, that of Barrow's Golden Eye being much shorter and higher at the base. Another method of distinguishing these birds is given at the end of the description of the plumage; but the dress of the females is almost identically the same. The present species frequents our lakes and rivers and feeds upon shell-fish and grasses. I have found it at times quite numerous on the St. Lawrence near Ogdensburgh, and have killed a goodly number there over decoys, and some specimens, procured on these occasions, are now in the Museum of Natural History in New York. The two species were associated together on the river, and I never knew which one would come to the decoys, but I do not remember that both ever came together unless it might be the females, for, as I have said, it was difficult to distinguish them without an examination.

The birds would fly up and down the river, doubtless coming from, and going to, Lake Erie, stopping occasionally in the coves to feed, and floating down with the current for a considerable distance, when they would rise and fly up stream again. My decoys were always placed in some cove or bend of the stream where the current was least strong, for I noticed the birds rarely settled on the water where it was running swiftly. This Duck decoys readily in such situations, and will come right in, and if permitted settle among the wooden counterfeits. They sit lightly upon the water and rise at once without effort or much splashing. The flight is very rapid, and is accompanied with the same whistling of the wings so noticeable in the Common Golden Eye. In stormy weather this bird keeps close to the banks, seeking shelter from the winds. It dives as expertly as its relative, and frequently remains under water for a considerable time. The flesh of those killed upon the river was tender and of good flavor, fish evidently not having figured much as an article of their diet.

CLANGULA ISLANDICA.

Geographical Distribution.—North America, from the Arctic regions south to northern New York, Illinois, Utah, and Colorado. Greenland, Iceland. Occasional in Alaska. Accidental in Europe. Breeds from Gulf of St. Lawrence, northward.

Adult Male.—A slight occipital crest. Head and upper part of neck, glossy bluish black, in some specimens with greenish reflections in certain lights. A large triangular shaped white patch, similar in form to a crescent, upper end pointed, lower end rounded, occupies the space at base of bill. Lower part of neck, and under parts, pure white; upper parts, velvety black; outer row of scapulars, with oblong white spots. A lengthened white patch on wing, formed by the middle coverts, ends of the greater coverts, and exposed parts of inner secondaries. Bases of greater wing coverts, black, forming a bar across the white portion. Feathers of sides and flanks, white with outer edges black. Thighs and sides of crissum, dull black. Tail, brownish black, with a greenish gloss. Bill, black. Feet and legs, orange yellow; webs, dusky. Length, about 22 inches; wing, 9, tarsus, $1\frac{1}{2}$; culmen, $1\frac{3}{16}$; height of bill at base, average, 1; width of nail, $\frac{1}{8}$; width of bill at base, $\frac{3}{4}$.

Adult Female.—Head and neck, snuff brown; darkest on top

of head and back of neck. A narrow white collar at base of neck. Upper parts, brownish black; feathers of back, margined with light gray. White patch on wing, crossed by a black bar. Lesser wing coverts, tipped with white. Upper part of breast, sides, and flanks, blue gray; feathers, edged with grayish white. Rest of under parts, white. Bill, horn color, paler in some specimens than in others, at times almost verging into yellow, with a spot on the culmen, and the edge of maxilla, and the nail, black or brownish black. Legs and feet, pale orange; webs, dusky. Wing, $8\frac{4}{10}$ inches; culmen, $6\frac{1}{2}$; height of bill at base, $1\frac{9}{10}$; width at base, $\frac{7}{10}$; width of nail, $\frac{8}{10}$; tarsus, $1\frac{4}{10}$. As a rule the bill of the female of this species is much shorter and higher at the base for its relative length, as well as narrower when viewed from above, than is that of the female of the Golden Eye. Mr. Ridgway distinguishes the two species by the height of the maxilla as compared with the distance from the feathered edge at the base of the bill to the farthest or anterior edge of the nostril. *In this species these measurements would be* EQUAL. Whether this character would hold good in a large series of specimens, I am unable to say, for there is a great variation in the size of the bills, but generally, I believe that this method of separating the females of these species can be relied on.

Downy Young.—Top and sides of head, chocolate brown, darkest on head. Neck, chest, and flanks, pale brown. Throat and under parts, white.

BUFFLE HEAD DUCK.

STRICTLY a North American species, the Buffle Head is found pretty generally from the Arctic Sea to Mexico and Cuba. In Alaska it is not common on the coast, but has been met with on some of the Aleutian Islands, and Stejneger procured it on the Commander Islands, showing that it goes to the Asiatic side of the Pacific Ocean, but probably only incidentally, not as a regular visitant. While a constant dweller upon our lakes and rivers, the Dipper also comes to the sea-coasts as soon as the inland waters are frozen. It is a cold-weather Duck, and only appears within our borders when driven south by the coming of winter. Its appearance is generally an indication that severe weather will follow. The male is one of the most beautiful and sprightly of our native Ducks, and is a great ornament to our waters. The female, on the contrary, in her subdued grayish brown plumage, lacks entirely the attractive dress of her lord, and would easily escape notice even among plainly attired birds. But the male seems conscious of his beauty, and, when upon the water, moves rapidly about, turning first one side then the other to the observer, and elevating or contracting his fluffy crest, causing its metallic colors to scintillate in the sun's rays.

This pretty Duck has had many names given to it among which are, Butter Ball, Spirit Duck, Marionette, Butter Box, and Scotch Teal; but those most generally employed are Buffle Head and Dipper, already mentioned. This species nests in hollow trees, lining the

bottom of the cavity with down, on which are deposited from six to ten grayish white eggs, and sometimes these are placed so far down in the hollow as to be almost out of reach, being more than an arm's length away from the opening. As a diver the Butter Ball takes rank among the most expert of our Ducks, disappearing so quickly, and apparently with so little exertion, that it is almost impossible to shoot it when sitting on the water. When alarmed, with a sudden flip up of its tail and a scattering of a few drops of water, it vanishes beneath the surface, appearing almost immediately at no great distance from where it went under, and either dives again at once, or takes wing, which it does easily and without any fuss. Sometimes half a dozen of these birds will gather together in a sheltered piece of water, and be very busy feeding. A few will dive with a sudden jerk, as if drawn beneath the surface by an invisible string, and the others will quietly swim about as if on the watch. The first that went under water having returned to the surface, the others dive, and so it goes on for a long time. Occasionally all will disappear, and then the first one to rise seems much disconcerted at not finding anyone on watch and acts as if he was saying to himself that if he " had only known their unprotected state, he would never have gone under."

The flight of the Buffle Head is very rapid, and generally performed in a straight line. So speedy is its course that it flashes by one like a feathery meteor, its wings forming a haze around the body, so quickly do they move, and it is no easy thing to kill one in the air as it hurls itself along. When alighting the bird makes a considerable splash and noise, sliding along for a couple of feet or so, before becoming stationary. It utters at times a single guttural note, which sounds like a small edition of the hoarse roll of the Canvas Back and other large diving Ducks. The male Dipper has a habit, when swimming, of stretching out and drawing in its neck, occasionally raising its bill as high as it possibly can, at the same time puffing out the feathers of the head. I have noticed that this is done mostly in the spring, when its thoughts are perhaps "lightly turning toward love," and it may be an attractive gesture common to the courting season. At all events, when the head is held high in the air, with crest expanded and the sun shining on its brilliant coloring, he presents for so small a creature a very gallant and handsome appearance. The flesh of this Duck is very palatable, and is excellent when broiled. In the spring the males precede the females on their northern migration, and arrive at their destination several days before their fair ones. The Dipper feeds on a variety of objects, such as fish and mollusks on the sea-coasts, and snails, leeches, grasses, and other water plants in the interior.

CHARITONETTA ALBEOLA.

Geographical Distribution.—North America, from the Arctic Ocean to Mexico and Cuba. Breeds from Maine and Montana, northward.

Adult Male.—A broad white band extends from behind and beneath the eye to the occiput. Rest of head and top of neck, glossed with metallic green, purple, violet, and bronze reflections. The feathers of the head are puffed out, and lengthened on sides and back. Lower part of neck, entire under parts, large patch on wing, composed of wing coverts and outer webs of secondaries, and scapulars, white. Inner secondaries, black. Primaries, black. Back and rump, black, fading into the pearl gray of the upper tail coverts. Tail, dark gray, with white edges to the feathers. Bill, bluish gray; nail, dusky. Iris, dark brown. Legs and feet, flesh color; webs, darker. Total length, about $14\frac{1}{2}$ inches; wing, $6\frac{1}{2}$; culmen, $1\frac{8}{10}$; tarsus, $1\frac{2}{10}$.

Adult Female.—Head and neck, dusky brown; top of head, darkest. A white patch or stripe on cheeks and ear coverts. Upper parts, blackish brown, grading into black on the rump. Wings, dusky brown. Apical half of outer webs of secondaries, white, forming a speculum. Upper part of breast, sides, anal region, and lower tail coverts, dull gray. Rest of under parts, white. Tail, grayish brown. Bill, dusky, slightly plumbeous on edge and tip. Legs and feet, bluish gray; webs, dusky. Total length, about $13\tfrac{1}{2}$ inches; wing, $5\tfrac{9}{10}$; culmen, 1; tarsus, $1\tfrac{1}{5}$.

The females vary slightly from each other, some having more white on the wings; the secondaries, and the tips of the greater wing coverts, also, being of that hue.

LONG-TAILED DUCK: OLD SQUAW.

IN North America the Old Squaw is found from the Arctic Sea to the Potomac and Ohio rivers and occasionally in Florida, Texas, and California, but it is met with mainly along the sea-coast, although in winter it is observed in considerable numbers on Lake Michigan and in Wisconsin. It comes to its far northern breeding grounds, on the Alaskan coast of Behring Sea, about the middle of May, being among the very earliest arrivals of the Duck tribe. It is found on the Aleutian Islands and has been known to winter around Unalaska.

Nelson states that these birds do not reach their nesting grounds from the sea until the ice has nearly all disappeared from the ponds and creeks, and the females begin to lay about the 12th of May, and from that date to the 25th. The nests are usually placed upon the sloping grassy banks of the ponds close to the water, and the parents keep in the neighborhood. During the period of courtship the male frequently swims rapidly about the female, with his long tail feathers elevated and vibrating from side to side, and during this display he utters his love note. The voice of this Duck is soft and with rather a sweet tone; the three notes usually uttered resembling somewhat the words, "Old, Soŭth Soŭthĕrly," or "Sŏuth Sŏuth Soŭthĕrly," ending with a rising inflection. Occasionally the female, when pressed by too ardent a lover, suddenly dives, followed by her partner, and they as quickly appear again and are on the wing, when a chase follows, both birds diving when at full

45. Long-Tailed Duck, Summer plumage.

speed, and mounting again in the air. This is kept up until both are tired. Occasionally other males join in the pursuit after the female, uttering their musical notes, until the lady, finding that she has too much company, retires to some secluded pond with her accepted lover, leaving the others to seek pastures new. In their habit of diving when on the wing during courtship without relaxing their speed, they are imitated by no other Duck save the Sprigtail. The nest is composed of grass stems and is lined with down, and the eggs, of an olive or grayish green color, are from five to nine in number. By the last of June the young are nearly all hatched and they remain about the ponds until the middle of August, when they usually go to the shores of the bays. It is one of the last species to leave the Arctic regions in the autumn, and does not depart until the ponds and creeks, and even the sea itself, are frozen over. In certain places, as some of the Aleutian Islands, where the sea may remain open at least to a considerable extent, it stays all winter.

The summer dress of this Duck is quite different from that of the winter, and is almost a sooty black with a rufous tinge upon the head, neck, and breast; the latter, however, being rather lighter. Sometimes, however, the winter dress, according to Nelson, is retained throughout the nesting season, and there is so much gradation observable among individuals between the two costumes that it is very difficult to procure any in perfect summer dress. As the ice commences to form the birds retreat, and get well out to sea before they begin their migration southward. It is, however, such a hardy bird, and seems so to love a freezing temperature, that it does not hurry, and goes on its way toward the south only as the waters become congealed or blocked with floes of ice, and thus compel it to move on. The Old Squaw breeds in Iceland and other parts of Northern Europe, also on the lower Anderson River, on the Barren Grounds, and on small islands in the bays on the Arctic coast. The number of eggs varies from five to seven, and they are always covered by the down plucked from the breast of the female.

This species does not seem to visit our Western coast, south of Alaska, but in its migration trends to the eastward, and enters our limits east of the Mississippi River. While it cannot be at all classed with the fresh-water Ducks, it is abundant at times on some of the larger Western lakes, making its appearance toward the last of October, about the time when all the smaller lakes and streams are frozen. It is fond of the sea, and is frequently seen in flocks off shore just beyond the line of breakers that hurl their white crests along the beach, rising and falling with the waves, or diving into the depths in search of food, or flying up and down parallel with the land, now disappearing between the billows in the trough of the waves, again rising above their crests, the flocks speeding on in long drawn out lines. The flight of this Duck is exceedingly rapid, indeed it may be regarded as among the swiftest of the tribe, and its powers of diving are excelled by none. It is so expert at this, and disappears from view so instantaneously and with so little effort, that it is next to impossible to kill it when on the water, the bird vanishing before the shot can reach it.

When the water is calm, and the sun has gained a certain amount of power as it returns on its northern journey from below the equator, the Old Squaws gather together in small parties on the open water of the sounds away from land, or on the bosom of the ocean, a gunshot or so from shore, and sleep or dress their feathers,

46. Long-Tailed Duck, Winter plumage.

perhaps dive a little and bring up some choice eatable from or near the bottom. At such times their musical notes are constantly borne to the observer's ear, of *Sŏuth, sŏuth, sōūtherly*, or as Nelson writes it *Â-leedle-â* (which, however, does not convey the sounds to my ear), and the pleasing chorus, rising from one portion or another of the assembled birds, disturbs with tuneful sounds the stillness that rests upon the sleeping water. The food of the Old Squaw is various small shell-fish, fry, and insects, fresh-water or marine, according to the locality in which the bird happens to be. As an article of food little can be said in favor of this Duck, for either the flesh is tasteless and tough, or else fishy and disagreeable. It is a very difficult bird to kill, for it flies with such swiftness that it is no easy mark to hit, and requires a very powerful blow to bring it down. If only wounded it is almost impossible to capture, as it dives with such dexterity and so persistently, and stays under water so long, that it will tire completely either man or dog.

In various parts of the land, besides those already given, and by which it is best known, it has many names, some of which are, South Southerly—from its cry, Old Wife, Old Injun, Old Molly, Old Granny, Cockawee or Caccàwee, Coween, Swallow-tailed Duck, Long-tailed Duck, Scolder, and Noisy Duck. There are others which, however, are mostly purely local, and familiar only to a few. The male is a handsome bird, whether in winter or summer dress, the long tail feathers being very ornamental, and especially conspicuous when the bird is rushing, with far more than the swiftest railroad speed, through the air. It is probably one of the species of Ducks that will remain with us the longest, as the poor quality of its flesh prevents it from being sought after as an article of food, and sportsmen pay little or no attention to it, save when no other Wild Fowl can be procured.

Perhaps, when, from continuous and ruthless slaughter, beginning with the destruction of the eggs in the far north, and the persecution of the birds throughout their long journeys to the southland and back to their breeding places, the majority of our Ducks have been annihilated, and the now despised Mud Hen or Blue Peter (*Fulica americana*) has become the game water bird of our successors, then the Old Squaw, in its descendants, may rise to the first rank of desirable Ducks, and be the choicest and most eagerly sought species of Water Fowl in the opinion of future sportsmen. But when that day comes, as undoubtedly it surely will, and the majority of our magnificent Water Fowl has become extinct, one dreads to think of the loneliness and stillness of our marshes, lakes, and tidal waters, which, once resounding in spring and autumn, aye and in many places throughout the winter, with the glad cries and cheerful calls of countless busy feathered creatures, will then lie tenantless and deserted, never more to echo with the voices of Nature's happy children, stilled forever.

HAVELDA GLACIALIS.

Geographical Distribution.—Northern Hemisphere. In North America from Arctic Ocean to Florida, Texas, and California, rare though in these States. In the Old World from Great Britain to Japan and China, occasionally in winter going to the Mediterranean. Breeds in Arctic regions.

Adult Male in Summer.—Lores, fore part of cheeks, and sides of forehead, mouse gray. A line above the gray from forehead passing over the eye and joining one from beneath the eye and extending to above ear coverts, white. Rest of head, neck, and upper parts, sooty black; the feathers on upper part of back

* See Appendix, page 290.

and the scapulars having the edges fulvous. Wing coverts, brownish black; secondaries, grayish on outer web, edged with whitish. Primaries, black at tip; dark purplish brown on outer webs, with light edges. Four middle tail feathers, black, with white shafts, the central pair greatly elongated, rest of tail, white; some feathers, dark brown on outer web along the shaft. Breast and upper part of abdomen, chocolate brown; rest of under parts and flanks, white. Bill, black, with a broad rose pink band crossing the maxilla in front of the nostrils. Iris, light hazel. Legs and feet, pale bluish white; webs, dusky. Total length varies greatly according to the elongation of the central pair of tail feathers, from 21 to 23 inches; wing, $8\frac{9}{10}$; culmen, $1\frac{1}{10}$; tarsus, $1\frac{1}{8}$; middle tail feathers, 8 to $9\frac{2}{10}$.

Adult Female in Summer.—Head and neck, dark grayish brown; space around the eye, and one on each side of neck, grayish white. Upper parts, blackish brown; feathers of upper back, with light brown tips; the scapulars almost entirely light brown, with blackish brown centers. Wings, similar to male. Upper tail coverts, blackish brown, feathers tipped with light brown. Tail, median pair not elongated, dark brown in center, growing lighter toward outer feathers, which are almost entirely white. Upper part of breast and anterior part of sides, light brown, rest of under parts, pure white. Bill, dusky olive gray. Legs and feet, bluish gray. Iris, yellow. Total length, about 18 inches; wing, $8\frac{1}{4}$; culmen, 1; tarsus, $1\frac{2}{10}$.

Adult Male in Winter.—Sides of head and orbital region, and in some specimens also, the lores, mouse gray. A large patch on sides of neck, black, grading into mouse gray on its lower portion. Rest of head, *including the eyelids*, and neck, upper parts of back and chest, white. Middle of back, rump, upper tail coverts, and wings, black. Scapulars, pearl gray. Secondaries, reddish brown. Tail, black on median feathers; central pair, elongated, growing lighter toward outer feathers, which are nearly all white. Breast and upper part of abdomen chocolate brown, in some specimens, black; rest of under parts, pure white. Bill, orange yellow; basal half on sides, and nail, black. Iris, carmine. Legs and feet, bluish gray.

Adult Female in Winter.—Forehead and crown, dusky; ear coverts, throat, and space about the eye, grayish white. Rest of head, neck, and lower parts, white. Jugulum, brown. Upper parts, dark brown; the scapulars, wing coverts, outer web of secondaries and feathers of the rump, edged with pale raw umber brown, sometimes with ashy. Tail, grayish brown, edges of feathers, ashy; central pair not elongated.

Young.—Similar to female, but the head and upper parts, darker and without the light border to the feathers of the latter. Lores, grayish brown, and the light patch about the eye smaller and rather indistinct; upper part of breast, brownish black, with gray tips to the feathers, graduating into the pale gray of the lower breast. Under parts, pure white. Feathers of the tail, grayish brown, with white margins. In this stage of plumage, this bird is very somber and unattractive.

Downy Young.—Head and upper parts, hair brown. Grayish white markings near eye; dusky stripe from corner of mouth to back of head. Under parts, white; dark brown band across breast.

47. Harlequin Duck.

HARLEQUIN DUCK.

AS fantastically decorated with various stripes as is the face of the Harlequin marked for the pantomime, this bird must rank as one of our beautiful species of Ducks. It is a native of the northern portions of both the New and Old World, and in the Eastern Hemisphere goes to Japan. In North America it ranges from the Arctic regions southward to the middle States and California, and breeds in the West from the Rocky and Sierra Nevada Mountains, and in the East from Newfoundland, northward. It cannot be said to be a common species anywhere, and few sportsmen have ever seen it in life. It is a solitary bird, except under especial circumstances, and goes either alone or in pairs, and haunts the most retired spots along the mountain streams, where the Ouzel delights to sport itself in the running water, or under the sparkling curtain of the foaming cascade. It breeds in such situations, but just where the nest is situated does not yet seem to be fully established. I have never seen it, and the accounts given of its situation prove that the bird alters its habits in a way not imitated by any other Duck, and influenced by the locality in which it may find itself. Thus Mr. C. W. Shepard states that he found it breeding in Iceland in holes in trees on the banks of the River Laxa, and Dresser says that the nest is placed on the ground, although he has never seen the nest himself, nor does he give any authority for his statement. Many observers have met with the old birds and their broods of different ages, but no one save Mr. Shepard and Mr. Pearson, hereafter mentioned, appears really to have found the nest. At one time the Harlequin Duck was not at all uncommon in winter on the Atlantic coast as far as New York, but of late years it does not come much farther south than the shores of Maine. In Alaska this Duck appears to breed in the interior along the mountain streams that flow into the great rivers, in the loneliest parts of that remote northern wilderness. The species is also at times quite numerous about the Aleutian Islands, frequenting the inner bays near the mouths of fresh-water streams, also in the outer bays and between the islands.

Nelson says that at the beginning of June at Unalaska the birds had united in very large flocks, of several hundred, were very shy, and when alarmed moved away with a confused noise of gabbling, chattering notes. He thinks they undoubtedly breed among the islands, but no nest was found. At the Seal Islands they remain all the year except when the ice compels them to leave for a season. The Indians along the Yukon stuff the skin of this Duck and decorate it with beads and bright cloth as toys for the children. The Harlequin Duck follows the West coast south as far as Puget Sound. It breeds in the Rocky Mountains at various altitudes, and according to Mr. Belden, as given by Brewer, he has seen numerous broods on the Stanislaus River, Calaveras Co., California, every summer at a height of about four thousand feet. The ducklings were exceedingly active in the water, tumbling over cascades and through rapids in a most astonishing manner.

Along our Eastern coast, from the Gulf of St. Lawrence to Maine, this Duck appears in greater or less numbers every winter. The females outnumber the males considerably, and it is easier to get a half dozen females or young birds than one male in perfect dress. It flies very swiftly, and when shot will often dive headlong into the water. In swimming the Harlequin sits lightly on the water, and the little flock (perhaps all members of one family), are usually preceded by the male, the others following demurely after him. If alarmed, they dive at once, and are very expert in all under-water tactics. This species feeds chiefly on mollusks and other shell-fish. Turner says that the common black mussel in Alaska is much sought after by this Duck, and it is constantly diving for it. Sometimes this bivalve seizes the bird by the bill, and does not release its hold until its victim ceases to struggle and so indicates that life is extinct.

As a rule the Harlequin is a silent bird, but in the mating season it utters a peculiar whistle, generally made by the male in his efforts to secure a mate. On account of its restricted range not many names have been applied to it, but it is known as Painted Duck, Rock Duck, and Lord and Lady, the latter on account of its beauty. In reference to the nesting and breeding habits of the Harlequin Duck, I wrote to my friend the late Captain Charles Bendire of the National Museum, Washington, who knew more about nests and eggs than any other man in America, and he replied as follows: "The Harlequin Duck undoubtedly nests both in our mountain ranges in the interior, Rockies, and Sierra Nevadas, as well as on many of the treeless islands of the Alaskan Peninsula and the Kurile Islands, and I have not the least doubt that it breeds both in hollow trees where such are available, and either on the ground or in holes made by Puffins where it can find such, not far from water. From what I have been able to learn from one of my correspondents I believe they breed early, even in Alaska. He writes me,

'I have killed many of them on the Kuriles during the months of May, June, and July, but they never contained ova of any size, so I conclude that they must lay earlier, and my belief has been strengthened by killing a female in Alaska which contained eggs as large as grapes early in March.' Mr. A. W. Anthony [continues Captain Bendire], writes me that a family of downy young were seen near Silverton, Colorado, on July 15th and one was taken. He states they are not uncommon there during the nesting season. They have also been observed during this time in Calaveras Co. and I have personally seen a family of eight or nine, with full-grown young in July, 1879, near Wenatchee, Kittitas Co., Washington, on the Upper Columbia, and shot two of the birds. There are no North American eggs of this species in the National Museum collection, and I do not believe its nest has as yet been found within the United States. I should judge the egg to be correctly described; it is figured by Hewitson in British Zoölogy, and by Baedeker, *Die Eier eurpaieschen Vögel.*"

In the *Ibis* for April, 1895, the Messrs. Pearson, writing upon some "Birds observed in Iceland," state that Mr. H. J. Pearson on the 11th of July, 1894, visited some islands composed of lava, in the middle of a river, and that the water ran like a mill race through three or four channels worn in the lava. On these islands he found six nests of the Harlequin Duck, three of them not two feet from the water hidden under the leaves of the wild anchelica, and the other three in holes in the banks, protected by a screen of plants. One contained seven eggs. Very little down was in any of the nests. Many old nests were in these holes, they having been apparently a favorite breeding place for years. Mr. Pearson saw flocks of more than thirty males together on several occasions sitting on the rocks, or sporting in rapids so swift that few birds would be apt to frequent them.

HISTRIONICUS HISTRIONICUS.

Geographical Distribution.—Northern portions of New and Old Worlds, ranging as far to the eastward as Japan. In North America from the Arctic regions to the Middle States and California. Breeds from Newfoundland and northern Rocky Mountains and Sierra Nevada, northward.

Adult Male.—Lores, with a stripe extending along the crown; a round spot near the ears, a long narrow stripe on side of upper hind neck, a narrow collar around lower part of neck, frequently interrupted in front; a broad bar across sides of breast in front of wing; middle of scapulars, portion of tertials, a round spot on lesser wing coverts, tips of some of the greater wing coverts and a round spot on each side of crissum, pure white. Under side of neck, collar, and bar on side of breast above and below, bordered with black. Center of forehead, crown, and nape, black, bordered on either side with chestnut. Rest of head and neck, dark plumbeous, glossed with violet, inclining to black along the margins of the white markings. Upper parts, leaden blue, grading into blue black on lower part of rump and upper tail coverts. Wing coverts, bluish slate. Speculum, deep bluish violet. Primaries and tail feathers, dusky black. Breast, plumbeous; abdomen, sooty gray, grading into the black of the crissum and under tail coverts. Sides and flanks, bright rufous. Bill, bluish gray in front of nostrils, basal part, dark olive gray, tip, paler. Iris, reddish brown or dark hazel. Legs and feet, bluish gray; webs, dusky. Total length, about $17\frac{1}{2}$ inches; wing, $7\frac{5}{10}$; culmen, 1; tarsus, $1\frac{4}{10}$.

Adult Female.—Lores, spot above and in front of eye, and larger one behind ear, white. Rest of head, neck, jugulum, and upper parts, dark brown, inclining to sooty brown on head and rump. Wings and tail, glossy blackish brown, with an inclination to a purple gloss in some lights. Breast, sides, flanks, crissum, and under tail coverts, light reddish brown, with indistinct black spots in the center of feathers on breast. Abdomen, white, becoming much mottled with brown on lower part, and passing into the reddish brown of the crissum. Bill, legs, and feet, dark bluish gray. Iris, brown. Total length, about 17 inches; wing, $7\frac{4}{10}$; culmen, 1; tarsus, $1\frac{1}{8}$.

The male in summer has a much duller plumage than in winter, and the pattern of the coloration not so clearly and sharply defined. In some portions of his dress at this season he resembles the adult female, and is not such a brave-looking gallant as he appears in the winter garb.

Young Male.—White markings of head and neck, less pure than in the adult, and the bar alongside the black on crown is yellowish brown, somewhat mixed with white. Head and neck, dusky, with a bluish tinge. Back and wings, dusky; edges of feathers, paler. Some of the tertials with white centers. No speculum. Tips of greater coverts, pale grayish brown, forming bar on the wing. Rump, sooty brown. Upper tail coverts, sooty, tipped with pale brown. Tail, light brown. The white collar at base of neck merely indicated, and the white bars before the bend of the wing about half as long as in the adult. Breast and under parts, sooty brown, mottled with white, lightest (almost white) on lower breast, and becoming reddish brown on under tail coverts. Flanks, pale chestnut.

Young.—Resembles the female, but darker above; the upper part of breast, sides, flanks, and under tail coverts, tinged with brown.

Downy Young.—Top of head and nape, blackish brown; cheeks and neck, white; upper parts, blackish brown; a white spot on each wing and thigh; under parts, white.

48. Surf Scoter.

SURF SCOTER.

THIS Coot is peculiar to North America and is found from the Arctic Sea to Lower California on the Pacific, and to Florida on the Atlantic coast. It is also met with on the Great Lakes, and through Illinois in winter, to Missouri. In fact its dispersion is almost precisely that of the White-winged Scoter. It breeds in similar latitudes, from Labrador to the Arctic Ocean on the eastern part of the continent, and at Sitka, also at the mouths of the Yukon, and about St. Michael's on the western side. As it goes in summer to both sides of Behring Straits, and to Norton and Kotzebue sounds, it may have other breeding places farther north than those given. In winter it is met with throughout the Aleutian Islands. The nest is similar to that of the White-winged Scoter, and is placed in like situations. The eggs, usually from five to eight in number, are white with a pinkish tinge. Sometimes in the far north males of this species collect together in immense numbers, and Nelson tells of a flock met with by him near Stewart Island, about ten miles out to sea from St. Michael's, which formed a continuous band around the outer end of the island for about ten miles in length and from one-half to three-fourths of a mile in width. As he drew near to this great mass the birds close to him began to rise, and their movements were imitated by those ahead of them until soon the entire mighty host of birds rose with the " roar of a cataract," and in a great black cloud swept out to sea, and settled again some distance away. Later in the season the females and young join these gay bachelors, and by the middle of October are met with in small flocks, all along the coasts, where they remain until the ice begins to form and drives them away. In the mating season they utter a low clear whistle, and will come to a decoy when this note is imitated. In the winter it frequents the sheltered coves and bays in the Aleutian Islands and is very shy and dives and goes a considerable distance under water, when alarmed.

The Surf Duck appears on our coast in company with the other Coots in October, and is the most numerous of all the species. It remains just outside the line of inner breakers, or between them and the beach, often coming quite close in, and in small companies passes the day in riding the waves and exploring their depths. It often enters the large bays, and occasionally is very abundant on the Chesapeake in the vicinity of Norfolk, and out toward the ocean. The birds are frequently seen dotting the surface of the water in every direction, and when a boat approaches, will wait until it is almost on to them, when they either dive, or rise heavily, flapping the water with both wings and feet until, gaining headway, they fly low for a short distance and drop with a splash into the waves again. All three species of Coots are often seen in such situations, but as a rule each keeps by itself, though occasionally, from the rapid approach of some steamer, the members of the flocks, on rising, get mixed together. But they do not remain so long, each species again seeking its fellows. As these Ducks are heavy and rise from the water with difficulty, they are always obliged to take wing against the wind, but if they are so situated that, to do this, they must fly toward the object of their alarm, they always take refuge in diving; frequently passing completely under a steamer and appearing on the other side.

As the weather increases in severity during the winter the Surf Scoters move southward, coming gradually northward as spring approaches, and by the month of May they are well on their way toward their northern breeding grounds. This species has straggled south as far as Bermuda, and there are two records of its appearance in that island, and it has also occasionally been captured in Europe, but these are merely wanderers from their fellows and native land, blown off their route possibly by some storm. The Surf Scoter has many trivial names, and is known as the Hollow-billed Coot, Skunk Head Coot, Spectacle Coot, Spectacle Duck, Surf Duck, Horse Head Coot, Bay Coot, Butterboat-billed Coot, etc.; while the females and young are called Gray Coot and Brown Coot. Although none of the Coots can be called handsome Ducks, yet the peculiar markings of the head, and the bright coloring of the bill of the present species, almost entitles it to that epithet. As an article of food the Surf Scoter is not generally much sought after, as its flesh is tough and fishy, but Turner says that in Alaska those obtained among the Aleutian Islands were very good indeed, and if well prepared the flesh was free from all strong odors. I am inclined to think that perhaps the absence of dishes obtainable in more southern climes, and the presence of an appetite excited by much open-air exercise, had a great deal to do with this opinion, for in the United States few people care to dine on Coot.

ŒDEMIA PERSPICILLATA.

Geographical Distribution.—Northern North America, from the Atlantic to the Pacific Ocean, and on the large inland waters. Going south in winter to Florida on the east coast; the

Ohio River in the interior, and to Lower California in the west. Breeds in the Arctic regions. Accidental in Bermuda and in Europe.

Adult Male.—Triangular spot on forehead, with the point forward, occupying nearly all the space between the eyes, and another large one on nape, pointing downward, white. Entire rest of plumage, glossy black, lightest on under parts; no white on wing. Bill has the " upper mandible (*maxilla*), above at base, including nostrils, dull crimson, this changed to flame scarlet over the front of the mandible (*maxilla*); nail, cadmium yellow, narrowly edged anteriorly with lighter yellow, and sometimes posteriorly with light lavender; sides with large squarish patch of black at base, this separated from the black feathering above it by orange, and from the feathering behind by a narrower edge of crimson; beneath this black patch and in front of it as far as anterior edge of nostril, or thereabouts, continuously white, the remainder of the sides (anteriorly to the white portion), pure orange; lower mandible nail like its fellow above; back of this for a short distance, reddish flesh color, terminating irregularly in white, the white continued to the base, with more or less black on the naked skin between the rami; feet, outer sides of tarsus and toes, excepting inner toe, crimson; the inner side, with both sides of inner toe, orange chrome, deepening in part to orange vermilion; joints and other portions blotchily marked with black; webs, solidly black."— *Trumbull.* Iris, white; pupil, black. Total length, about 21 inches; wing, $9\frac{1}{4}$; culmen, $1\frac{1}{2}$; tarsus, $1\frac{7}{10}$.

Adult Female.—Top of head and nape, brownish black. A more or less distinct patch on lores, and another behind the ears, white. Rest of head and neck ashy brown. Upper parts, dusky brown, with some feathers having paler tips; under parts, grayish brown, nearly white on the abdomen; some of the flank feathers tipped with white or whitish brown, anal region and under tail coverts, dusky. Wing like the upper parts, no white. Bill, black, with a greenish tinge. Iris, dark. Feet and legs, brownish yellow; webs, black. Total length, about 19 inches; wing, 9; culmen, $1\frac{1}{5}$; tarsus, $1\frac{3}{10}$.

Young Male.—Resembles very closely the female, but the white spots on lores and sides of head are clearer, and there are traces, or beginnings, of the large white patch on the nape. Bill is slightly tumid at base, with pinkish tinge on sides anteriorly.

Among adult males there is considerable variation in the white marking of the head. Sometimes the patch on the fore part of crown is wanting, and there is considerable difference in the size of these markings when present. The coloring of the bill, also, varies at times from the typical style, some individuals having more black than others.

AMERICAN SCOTER.

THIS Duck has a wide dispersion in North America, and is found from the Arctic Sea to Southern California on the Pacific, to the Great Lakes in the interior of the continent, and to New Jersey and possibly much farther south on the Atlantic coast. It has been procured at St. Louis in Missouri, and is a rare visitor to Illinois and perhaps some of the adjacent States. It is abundant at Hudson Bay, but is present in greater numbers, in the breeding season, on the Alaskan coasts than in almost any other portion of the extreme north. It abounds about Behring Sea and Kotzebue Sound, and has been seen at St. Lawrence Island, breeds on the Nearer Islands, occurs on the Commander Islands and the Shumagin group, is a winter resident in the Aleutian Islands, and, according to Swinhoe, has been taken in China and Japan.

The species comes to St. Michael's, Nelson says, when the ice begins to break up on the sea, and the ponds in the marshes are open. Toward the end of May the birds frequent these last, and mating having been accomplished a site for the nest is chosen. This is generally in the grass near to water, and formed of grass, feathers, leaves, and moss. If any low-branching tree or bush is handy, the nest is often placed beneath it. As a rule it is carefully hidden, and the eggs are covered by the female whenever she goes away. When incubation commences the males leave the females and gather, as is the custom of the Eiders, in great flocks along the sea-shore

49. American Scoter.

in the vicinity of a bay or inlet. These assemblies continue to grow in numbers throughout the summer. Sometimes the males are seen with the females in the marshes throughout the season, but these are late breeders. The young are kept by the females near the nest in some pond until half grown, and then they gradually work their way down to the sea. Their habits during the breeding season are very much like those of the Eiders.

About the middle of October the migration southward begins. Upon the Atlantic coast the American Scoter appears from its northern breeding grounds in September. These individuals are mainly old birds, the young coming during October. They are associated with the two other species of Scoter, and continue to pass along the coast until late in the winter. The present species is less numerous than the others, and while the members of the flocks usually keep pretty well together, they yet at times become all mixed up with the White-winged and Surf Scoters. They keep at quite a distance from the beach, and fly in a long line just above the water, headed generally by some old male. They travel at a great speed, sometimes at the rate of, possibly, one hundred miles an hour, and are very difficult to kill, not only because of the rapidity of their flight, but also on account of the density of their feathers, which to a great extent prevents the shot from entering the body. As a diver, like all Sea Ducks, this Scoter is most skillful, disappearing without effort beneath the surface, and remaining for an exceedingly long time without rising. If wounded it will frequently seize some grass growing on the bottom, as already related of some other deep-water Ducks, and commit suicide by drowning rather than permit itself to be captured. If the water is clear, the bird can be seen close to the bottom, and if an oar can be made to reach it, by repeated pushes it can be compelled to release its hold, when it usually rises to the surface, though sometimes it will swim to another clump of grass and hang on to that.

This species utters a long musical whistle, and it can often be distinguished by this note from the other Ducks in the vicinity. In windy weather these birds fly very low over the water, and if disturbed by a passing boat, when resting on the surface, if they rise at all, it will be to fly for only a short distance, and then drop with a splash, and usually dive at once if the object of their alarm is near. In calm weather they fly very high, especially when migrating. They mate, as do many of the Water Fowl, before the spring migration begins, and the male will often be seen following the female closely about whenever she is upon the wing. Should anything happen to her he frequently returns to seek her, but if he is the sufferer she pays no attention to him, but continues on her way with apparent indifference.

This Duck has many names, the best known being, Black Coot, Whistling Coot, Butter-billed, and Hollow-billed Coot, while the female is called Gray and Brown Coot. There are quite a number of other names, many of them purely local. The color of the eggs is a pinkish ivory white. The male of this Duck, while arrayed in a melancholy dress of intense black, has one brilliant spot, in the place that would be most suspicious and unattractive in man, but which is all right in a bird, viz.: around and behind the nostrils. The basal part of the bill bulges up and is a bright orange, slightly paler above. This bit of color relieves the appearance of what would be otherwise a gloomy and somber-looking creature. As a bird for the table, the adults of this species, like those of the two succeeding, are abominable.

ŒDEMIA AMERICANA.

Geographical Distribution.—Northern North America, from the Arctic Ocean to California on the Pacific, and to the Great Lakes in the interior (accidental in Missouri), and to New Jersey on the Atlantic coast. Breeds from Labrador throughout the Arctic regions, Aleutian Islands, and Islands of Behring Sea, and is said to visit China and Japan.

Adult Male.—Entire plumage, black, glossy on head, neck, and upper parts. No speculum. Inner webs of primaries, grayish. Bill, black on apical half, bright orange on basal half, including the gibbous portion, or knob. Iris, deep brown. Legs and feet, blackish. Total length, about 18 inches; wing, $8\frac{1}{4}$; culmen, including knob, $1\frac{3}{4}$; tarsus, $1\frac{7}{10}$.

Adult Female.—Front, crown, and nape, dark brown. Chin, throat, and sides of head and neck, light grayish brown, speckled with dusky. Upper parts, sooty brown, tips of feathers, lighter: under parts, grayish brown; feathers on lower breast and abdomen, frequently tipped with grayish white. Bill of normal shape, black, sometimes with yellow marks. Legs and feet, olive brown; webs, black. Total length, about 18 inches; wing, $8\frac{1}{2}$; culmen, $1\frac{7}{10}$; tarsus, $1\frac{6}{10}$.

Young.—Resembles the female. Chin, throat, sides of head and neck, brownish white. Under parts, whitish, with nebulous spots of brown. Crissum, grayish brown; feathers, with whitish tips.

Downy Young.—Upper parts and breast, dark brown. Throat, white. Abdomen, grayish brown. Bill, dark plumbeous. Legs and feet, olive.

VELVET SCOTER.

THE Velvet Scoter is a bird of the Old World, and has only been obtained a few times within the limits of North America, viz., in Greenland and Alaska. It must then be regarded as an accidental visitor to our shores, and in no way considered as an American species. It is rather common in the northern portions of the Eastern Hemisphere, going southward, during the winter, to the Mediterranean and the Caspian seas. Like its American ally, this Scoter is found along the sea-coast, flies swiftly after it once gets started,—for it is rather clumsy in rising from the water, as it is a heavy bird,—swims easily, and is a most expert diver.

The nest is placed upon the ground near some pond. It is merely a depression, hidden under a bush, and lined with grass, leaves, and some down, and the number of eggs varies from eight to ten, ivory white in hue, with a buff tinge. The habits of this species are the same as those of the American Scoter. When incubation begins the males desert the females, and assembling together resort to the sea, and the islands lying off shore. When the young are full grown they and the females join the males, and begin their journey southward. The Velvet Scoter bears a close resemblance to the American Scoter, but has a differently shaped and colored bill, which easily distinguishes the two forms.

50. Velvet Scoter.

ŒDEMIA FUSCA.

Geographical Distribution.—Northern portions of the Old World. Occasional in Greenland and Alaska.

Adult Male.—General plumage, uniform velvety black. Eyelids and spot under eyes, white. Speculum formed by tips of greater coverts and secondaries, white. Bill, orange, much elevated at base, with a black line running obliquely from nostril to the nail. Iris, white. Legs and feet, dark red, or crimson; webs, black. Total length, about 22 inches; wing, 11; culmen, $1\frac{6}{10}$; tarsus, $1\frac{9}{10}$.

Adult Female.—A spot near base of maxilla, and one near the ear, and also the secondaries, white. General plumage, brownish gray, with pale edges on the back and scapulars. Under parts, sooty gray; feathers, edged with whitish. Bill, dusky. Legs and feet, similar in color to those of male, but paler.

Downy Young.—Resembles those of *Œ. americana*, but has a white spot on the wings, and the belly white.

WHITE-WINGED SCOTER.

A WELL-KNOWN species along our coasts, and on our inland lakes and rivers in certain portions of the West, the White-winged Scoter has a wide distribution throughout North America. It does not seem to breed as far north as the American Scoter, but has been obtained on both sides of the continent, and goes in winter on the Pacific coast as far as Southern California and to the Middle States on the Atlantic. It is also found upon the Great Lakes, being common on Lake Michigan, and is generally met with throughout Illinois in winter and has been seen in Wisconsin, Minnesota, and as far south as the vicinity of St. Louis, in Missouri.

As the White-winged Coot it is known everywhere, and is usually considered of little value, on account of the poor quality of its flesh. Its habits resemble those of the other Coots, with which it is frequently associated. In Alaska it breeds, about St. Michael's, on the lower Yukon, and also in the vicinity of Sitka, and occurs very sparingly among the Aleutian Islands, but in autumn is common along the coast of the mainland from St. Michael's, southward. On the eastern side of North America it breeds along the Mackenzie River to the Arctic Sea, on the Lower Anderson River, and on the Barren Grounds, and at Hudson Bay. The nest is placed upon the ground, concealed in a clump of trees, or under some low, spreading bush, and is a mere depression in the ground, lined with down and feathers, and near some

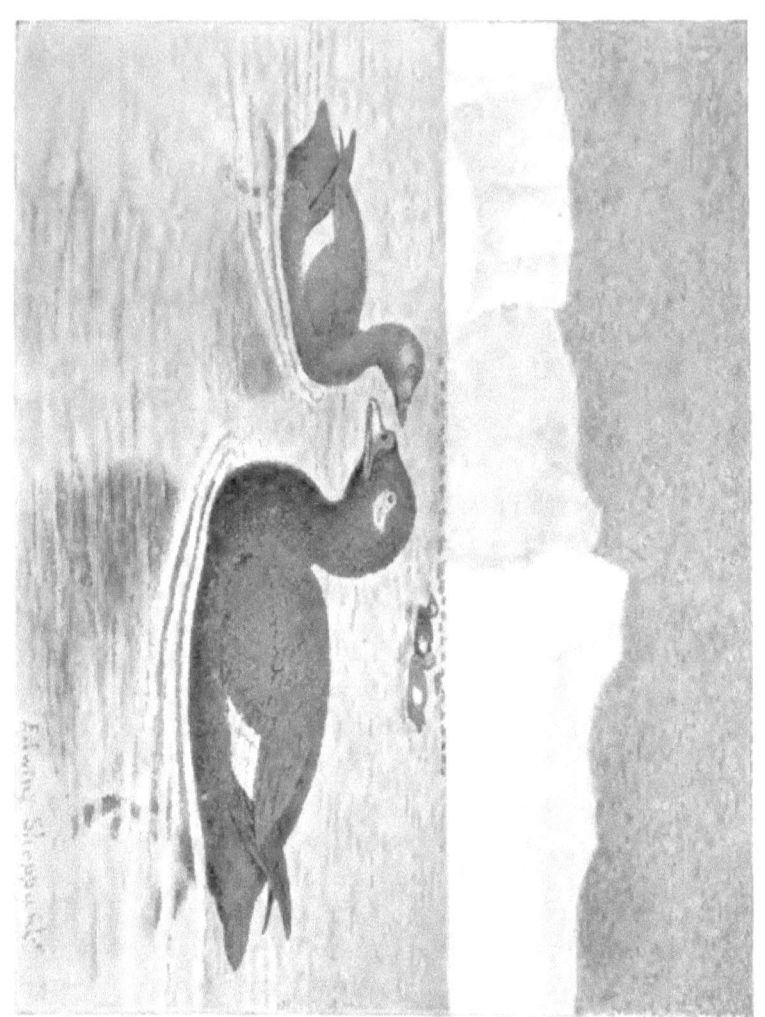

51. White-Winged Scoter.

pond or stream. The eggs, which are from five to eight in number, are a light cream color. This Duck breeds, in addition to the places already mentioned, in Labrador, where Audubon found the nests and eggs and also captured some of the young, only a few days old. The nests were placed in situations similar to those already described, but were formed of twigs, mosses, and plants matted together and without any down. He says the eggs he saw were pale cream color, tinged with green. Although the young he procured were only about a week old, the males could already be distinguished from the females by the white spot under the eye. The down covering them was stiff and hairy, all black except the chin, which was white. The birds were present in great numbers and kept arriving all the time from the Gulf of St. Lawrence.

On the Atlantic coast, this Coot reaches the shores of the New England States in September or beginning of October, appearing in flocks of no very great size, the old ones coming first. When migrating the birds fly high, and they pursue their way in silence. If the weather is stormy their course is low over the water, rising and falling with the waves, now just topping some combing billow, again hidden as they disappear in the trough of the sea. Although their flesh is poor, tough, and fishy, numbers of this Duck are shot by gunners every season. They are exceedingly tenacious of life, and are clothed in such a dense feathery covering that it requires a gun heavily charged to bring them down; and if only wounded they dive so quickly, and stay under water so long, that it is next to impossible to secure them. The feathers also, beside being strong and thick, seem as if they were inserted through the skin and clinched on the under side, and the labor of picking a few individuals of this Coot is no joke, usually resulting in sore fingers.

Where a flock is flying too high for a successful shot, it can sometimes be brought within range by discharging a gun at it, and at the report, the birds will often make a sudden plunge downward in the direction of the water, coming near enough to the sportsman for him to kill some of them. The Scoter, as has been said, is a skillful diver, and will frequently go to the bottom, where the water is fifty feet deep, and, if wounded, stay there.

This Duck has many names among the sportsmen and gunners, some of the best known being, Velvet Duck, Velvet Scoter, White-winged Surf Duck, Coot, Black Surf Duck, etc. It is also the Lake Huron Scoter described by Herbert (Frank Forrester) from an immature bird, and although the young, when it has fed perhaps upon such diet as the inland lakes afford, is not (as I know, for I have shot numbers of them in such waters myself) as fishy as the birds killed on the coasts, yet it does not deserve the praise he gives it. The food of this Duck consists of fish, mollusca, and various crustaceans procured by diving.

ŒDEMIA DEGLANDI.

Geographical Distribution.—Northern portions of North America on both the Atlantic and Pacific coasts; going south in winter to Chesapeake Bay, southern Illinois, vicinity of St. Louis in Missouri, and Lower California. Breeds in the Arctic regions.

Adult Male.—A small spot under eye, and speculum on wing, white. Entire plumage, black, inclining to brownish black beneath; flanks olive brown. Base of maxilla, including elevated culmen and nostrils, together with the edges, black. Sides of maxilla, deep red, grading into orange on culmen; nail, vermilion. Between the nail and nostril, pearly white. Iris, white, or yellowish white. Legs and feet, scarlet; joints and webs, black. Total length, about 20 inches; wings, 11; culmen, $1\frac{6}{10}$; tarsus, 2.

Adult Female.—Upper part of head and neck, dark brown. A white spot behind the ear, and another indicated at base of maxilla. Rest of head and neck, sooty brown. Upper parts, sooty brown; speculum, white. Under parts, grayish brown. Bill, uniform dusky. Iris, dark. Legs and feet, duller than those of the male, flesh color, tinged with black; webs, black. Wing, $10\frac{1}{4}$; culmen, $1\frac{8}{10}$; tarsus, $1\frac{7}{10}$.

Young Male.—Similar to the female, but head and neck, sooty black, and no white spot beneath the eye as in adult male.

STELLER'S DUCK.

THIS very beautiful bird is only found within our limits, on the coasts and islands of Behring Sea; it also breeds along the northern shores of Siberia, and from there occasionally straggles into Russia and northern Europe. About all we know of it is derived from the accounts of the few naturalists who have visited its habitat. Nelson says he found it rather numerous in the quiet, sheltered bays and fjords of the Aleutian Islands, where, however, it was very shy. The residents told him the species was abundant in winter in the bays not ice-bound, and many birds were killed for food. It is found at Kadiak and Sauk Island, near the eastern end of the Aleutian chain, on the Shumagin Group, and also in great flocks on the north coast of the Alaskan Peninsula. Dall found it associating with the King Eider in winter.

The mating season begins in May, and the nest is placed between tussocks of grass and lined with the same material, and concealed by long, overhanging grasses. In the nest found there was a single egg, which was a pale grayish green color. It is said that if a nest is visited the bird will abandon it at once.

This Duck seems to be irregular in its movements and does not always appear at the same place at stated periods. Thus in May, 1872, it was very abundant at Unalaska together with the Pacific Eider, but in May, 1873, though the season was later, not a single member of either species was seen. It breeds on St. Lawrence

52. Steller's Duck.

Island, and in summer passes along the Siberian coast from Kamchatka northward, moving, as winter approaches, to the Aleutian and Kurile islands to the south. Steller's Duck frequents lagoons and the mouths of large rivers, also outlying rocky islets and exposed reefs, and feeds in the tide-rips, keeping along the shore but not very close in, where the water is clear and deep. It dives well and remains a long time below the surface, seeking its marine food. In the Arctic Sea, at Point Barrow, this species arrives in June and leaves by the middle of August, and in Norton Sound the birds are taken as late as the 15th of October, but those captured are mostly young of the year. At the Commander Islands they arrive at the beginning of November, stay all winter, and leave by the end of May. It will be seen that this handsome Duck is a lover of cold weather and ice-bound coasts, and makes no effort to join the hosts that annually in the autumn seek a milder clime, but dwells throughout the year along the cheerless, sterile shores that bound the Arctic seas and islands. Although it has been taken in various parts of Northern Europe, even in the British Islands, there is no record of its capture in North America south of the limits of Alaska.

HENICONETTA STELLERI.

Geographical Distribution.—Arctic and Subarctic coasts of Northern Hemisphere, Islands in Behring Sea, Aleutian Islands, and coast of Alaskan Peninsula, east to Kadiak.

Adult Male.—Head and upper part of neck, satiny white; space around the eyes on each side of occipital tuft, chin and throat, black. Lores and tuft of stiff feathers on occiput, pale olive green. Lower part of neck, middle of back, rump, and upper tail coverts, blue black. Long scapulars, shining blue black on outer web, and white on the very narrow inner web;

other scapulars similarly colored, bend downward across the wing, falling below the primaries. Wing coverts, anterior scapulars, and sides of back, pure white. Outer webs of secondaries, shining blue black, forming a speculum, the tips white, making a bar below the blue black. Primaries, blackish brown. Under parts, tawny, deepest on breast and middle of abdomen, which are chestnut or even black, grading into light buff or ochraceous on sides. A round black spot on each side of the breast in front of primary coverts. Anal regions and under tail coverts, black. Tail, brownish black. Bill, light bluish gray, yellowish at tip. Iris, dark brown. Legs and feet, brownish gray. Total length, about 18 inches; wing, 8; culmen, 1½; tarsus, 1⅜.

Adult Female.—Head and neck, reddish brown, speckled with dusky. Upper parts, dusky; feathers, tipped with fulvous. Wings, dusky; the coverts, tipped with brownish gray; tertials, broadly margined with snuff brown. Tips of greater coverts and secondaries, white, forming two narrow bars across the wing. Primaries, blackish brown; speculum, dull purplish brown. Upper part of breast, rusty brown, spotted with black; upper parts, sooty brown. Bill, bluish gray. Legs and feet, brownish gray. Total length, 17½ inches; wing, 8; culmen, 1½; tarsus, 1⅜.

53. Spectacled Eider.

SPECTACLED EIDER.

A RESIDENT of the remote northwest coast of America, the Spectacled or Fischer's Eider, as it is sometimes called, is local in its habitat, and is met with from the mouth of the Kuskokwim River to Point Barrow, appearing at the latter place in summer. Its breeding range, according to Nelson, is from Norton Bay to the Kuskokwim River, but Turner says it also occurs among all the Aleutian Islands, where it breeds and is a constant resident, although extremely shy. This is another of our wild Ducks that have never appeared south of Alaska, and only those who have visited the extreme northern part of that Territory, above the Peninsula, have had any opportunity to observe it in its native haunts. Its dispersion is somewhat greater than was at first supposed, but, even as we now know it, the species appears to be very local. It arrives in the vicinity of St. Michael's between the middle and last of May, flying in small flocks not exceeding fifty individuals, and skimming just over the surface of the ice or marsh. Nelson says that the flocks break up soon after reaching their destination and mating takes place, but the eggs are seldom laid before June. The lovemaking is of a quiet, undemonstrative kind, and the birds are silent, uttering no notes. The nest is a depression amid the grass, in some dry spot near the water, and lined with grass. The eggs, from five to nine of which make a set, are light olive drab in color. Other nesting places are tussocks of grass, small islands in ponds, and knolls near the water, and the nest is hidden in the dry grass amid which it is placed. The male remains near the nesting place until the young are hatched, when he disappears, probably to moult, and the female takes sole charge of the young and shows much courage in their defense, putting herself in the way of danger, and shielding the little ones from harm by every means in her power. By the beginning of September the young are well grown and all have deserted the marshes, and the species is scarce along the coast toward the last of the month. Nelson thinks that on account of its local distribution, and restricted range, it may readily be so reduced in numbers as to become a very rare bird, possibly even extinct, like the Great Auk and Labrador Duck. Its breeding range does not exceed four hundred miles of coast line with a width of not over one or two miles, and against the usual opposing natural forces it must contend with, it has, in addition, the natives armed with shotguns. The diminution of Water Fowl in that country, he says, is more marked every season, and this in certain cases can only be the beginning of extinction, and this warfare against the feathered creatures will be increased on account of the growing scarcity of large game.

ARCTONETTA FISCHERI.

Geographical Distribution.—Alaskan coast from the Kuskokwim River to Point Barrow, Behring Sea, Aleutian Islands.

Adult Male.—Feathers projecting onto the bill, stiff, plush like, yellowish white, anteriorly grading into sea green on the lores and forehead, this color extending in a narrow line along the crown, and in a rather broad stripe beneath the eye patch, and then broadening out on the thick occipital crest. The green is deepest on lores, on the stripe under the eye and edges of crest, and becomes very pale yellowish green on crown and center of crest. A large satiny white pad encircles the eye and covers nearly all the side of the face and crown, bordered above and on either side by a narrow line of black. Chin, throat, neck, back, small wing coverts, scapulars, falcate tertials, and a large patch on each side of the rump, white. Greater wing coverts, primaries, and tail, dark brown. Under wing coverts, pale brown. Lower back, rump, upper tail coverts, and breast, dark plumbeous, grading into smoky black on the lower breast, abdomen, and under tail coverts. Bill, orange, deepest along the edges, and palest on nail. Iris is surrounded by a broad, bright, milky blue ring. Legs and feet, olive brown or yellowish. Total length, about $21\frac{1}{2}$ inches; wing, 11; culmen, $\frac{8}{10}$; tarsus, $1\frac{9}{10}$.

Adult Female.—Fore part, top, and back of head and back of neck, yellowish buff, streaked with dusky, coarsest on back of head and neck. A broad stripe, about $\frac{3}{8}$ inch in width, in front of eyes, beginning at corner of the mouth and extending onto center of head as far as posterior line of eye, dark brown. Space around eyes and cheeks, grayish buff, finely streaked with dusky. Upper parts, rather coarsely barred with fulvous and black, the bars narrower on rump and upper tail coverts. Lesser coverts of wing, pale brown, barred with black. Remainder of wing, pale brown; the tips of greater coverts and secondaries, white, forming two bars across the wing. Breast and sides and under tail coverts, barred with fulvous and black. Rest of under parts, grayish brown. Bill, dull blue. Legs and feet, dull yellowish brown. Total length, about 21 inches; wing, $10\frac{1}{2}$; culmen, 1; tarsus, $1\frac{3}{8}$.

AMERICAN EIDER.

REPLACING the Common Eider on a large portion of the Atlantic coast, the present species is distributed from Labrador as far south in winter as the Delaware River. Formerly it was more abundant and passed a greater portion of the winter along the shores of Massachusetts, but now it seldom appears south of that State save in very cold weather in midwinter. Occasionally it penetrates to the westward, and has been observed on the Great Lakes and captured in both Illinois and Wisconsin, pretty far in the interior for a Sea Duck. It breeds from the northern limit of Labrador to the Bay of Fundy and the northeastern coast of Maine. In Labrador it prefers small islands in sheltered bays as sites for its nest, and this is placed under small firs and other trees with low, down-reaching branches, or beneath overhanging plants with thick foliage. The nest, placed in a depression in the ground in situations like those described, is formed of sea-weed, mosses, grasses, and such-like suitable material, and filled with the softest and warmest downy bed imaginable, in which the eggs lie, often hidden from sight. These are usually six in number, pale greenish olive in color. The female is a close sitter, and if disturbed from the nest utters a hoarse croak. Sometimes one nest is occupied by two females, each depositing her eggs, and when the full number is reached both carry on the duty of incubation together in the most complete harmony, and when the young appear assume jointly the care of the united broods. The female defends her young from the attacks of such feathered and

furred foes as she is able to withstand, and, as soon as they are hatched, leads them to the water, where they can at least escape from their enemies of the air by diving. The males leave the females when incubation begins, and, like those of the other species, betake themselves to the sea. The food of this Duck consists of mollusks, which it swallows entire.

The American Eider likes to haunt rocky shores, and may often be seen standing on the bowlders, slippery with the spray and marine mosses, at the edge of the water. I have frequently watched them flying low over the sea in regular undulating lines, the quick flaps of the wings, succeeded by a rigid poise, when on fixed pinions the birds would sail along for a short distance, followed by more flappings, and thus, with alternate beats and sailings, they would move swiftly along close to the shore. Occasionally they would be congregated on the water in flocks of considerable size, and avoided the approaching boat by diving, staying under the surface for a rather lengthy time, and then rising at some distance away, to dive again, or to move off in long lines.

The male is a handsome bird, and shows well when swimming on the surface of the sea, as he rises and sinks upon the swells rolling in toward the rocks. As an article of food the American Eider is about on a par with his European relative, and there is little satisfaction in shooting the bird, large and handsome as he is, unless for the sole purpose of obtaining some down or a specimen.

SOMATERIA DRESSERI.

Geographical Distribution.—North America, from Labrador to Delaware on the Atlantic coast. Occasionally westward to the Great Lakes.

Adult Male.—In color of plumage and its general distribution there is no appreciable, certainly no specific difference, between the male of this species and the Common Eider of Greenland and the northern regions of the Old World, and the description given of the succeeding species may answer very well for the American Eider. But the two forms, apart from their plumage, can be readily distinguished by the shape of the frontal angles, or the naked portion running from the base of the bill onto each side of the forehead. In *S. mollissima*, the next species, these angles are narrow and more or less pointed and smooth, while in the American Eider they are broad and rounded at the end, and much corrugated. In general measurements there is very little difference between the two species. The bill of the present one, from tip to end of frontal angle, averages about $2\frac{1}{10}$ inches; greatest width of angle, .45; culmen, $1\frac{9}{10}$.

In some male specimens a dusky V-shaped mark is observable on the throat, but this is rare.

Adult Female.—With the exception of the shape of the frontal angle, the female of this species is not to be distinguished from that of the Common Eider.

Downy Young.—Like that of the Common Eider.

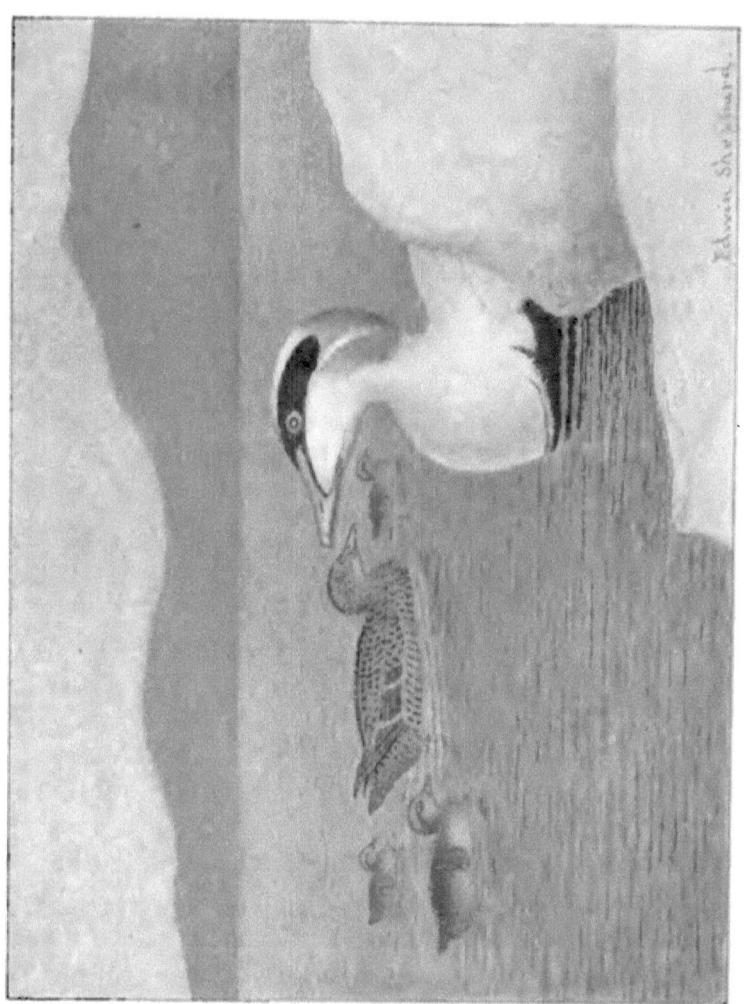

55. Eider.

EIDER.

IT was for a long time even unsuspected that there was more than one species of the Common Eider Duck, so well known throughout all the world for the valuable quality of its down. But when the birds from the Atlantic coasts of North America were critically compared with those from Europe, sufficient differences were discovered to necessitate the recognition of two distinct species. The plumage of the two forms does not differ, but the characters that separate them are found in that part of the maxilla, or upper half of the bill, which extends toward the head. In the present species, which is the same as the European bird, this portion of the bill is narrow and terminates in a point, while in the American species, this part is broad and has a convex end. The Common Eider ranges throughout northern Europe, and is found in Greenland and on the American coast from Labrador south in winter along the shores of Maine, and in the Arctic Ocean as far west as the Coppermine River in the longitude of Great Slave Lake. The down of this bird, which is plucked from the breast of the female for a lining to her nest, is a most valuable article of commerce, and in order to procure it in sufficient quantity, the birds may almost be said to have become domesticated in Iceland, Norway, and other parts of Europe, where they are in the habit of breeding in large numbers. Accommodations are provided for them, the turf is scraped away in squares of about eighteen inches each, or similar spaces are arranged with stones, all of which are occupied in the season by sitting birds, and so closely are the nests placed to each other that one can hardly move among them without stepping upon a Duck or an egg. In such colonies as these the Eider become very tame, and frequently will not leave the nest when a person approaches, and some allow the inhabitants of the island, whom they are accustomed to see daily, to stroke their feathers or remove the eggs from beneath them without more remonstrance than is usually made by a hen under similar circumstances. By the time the full complement of eggs is laid, the down has been gradually increased in the nest, until at length the quantity becomes so large that the eggs are entirely concealed and covered by it. The nests are made of sea-weed, and the eggs, five or six of which are a full complement, are a pale green color.

When incubation has commenced the males retire to the sea and remain in flocks near the shore, leading an idle, careless kind of a bachelor life, free from all family duties, and when moulting time arrives they go farther out to sea, and do not return to the females and young until the autumn. Incubation lasts about a month, and the young are conducted to the water by the female, sometimes carried there in her bill, and she remains with her little family until they are full grown and are joined by the males, later in the year. This Duck does not seem to mind cold, and has been known to endure a temperature of 50° below zero without any inconvenience. Of course it could remain in such extreme frost only in places where the water was kept open, and comparatively free from ice, by the rapidity of the current or tide rifts.

The Eider is a great diver and remains a long time under water. It feeds chiefly on mollusks which it procures on the bottom, often at great depths. The flight is low and performed in Indian file, each bird following at a regular distance from the one in front, and by regular flaps and sailings of the wings. The males make a sort of cooing sound, especially when sitting near the shore during the breeding season, and the females often leave the nests for a short time and join them. Although breeding, and not uncommon, in various parts of the eastern Arctic regions in North America, it cannot be said to appear often, at least in any considerable numbers, on our Atlantic coast much south of the Gulf of St. Lawrence, but is supplanted there by its near ally the American Eider. Both are large Ducks of about equal size, and on the wing it would be impossible to distinguish one from the other. As an article of food, the Eider cannot be said to take very high rank, but from the nature of its diet has a fishy, unattractive quality of flesh. The eggs are said, however, to be palatable.

SOMATERIA MOLLISSIMA.

Geographical Distribution.—Northeastern coast of North America, south to Massachusetts; Greenland, northern part of Eastern Hemisphere.

Adult Male.—Top of head, velvety black, with a white stripe in the center of the occipital region. Nape and posterior part of the auricular region, sea green; cheeks, neck, chin, throat, back, lesser and middle wing coverts, falcate tertials, and a large patch on either side of rump, pure white. Greater wing coverts and secondaries, brownish black. Primaries, pale brown. Lower part of back, rump, upper and under tail coverts, and entire under parts below the breast, deep black. Breast, pinkish cream color. Sometimes the back and scapulars are tinged with yellowish. Tail, pale brown, like the primaries. Bill, olive green; sometimes with an olive yellow shade; nail, greenish yellow. Legs and feet, olive green. Total length, about 22 inches; wing, 12; bill, culmen, 1_{10}^{9}; from tip to end of frontal angle, $2\frac{3}{4}$; greatest width of angle, 30; tarsus, 1_{10}^{8}.

Adult Female.—Head and neck, pale rufous brown, streaked with narrow black lines; upper parts of head, darkest. Rest of plumage, brownish buff, or chestnut brown, on the upper parts and breast, barred with black; the under parts below the breast, grayish brown, with dusky nebulous bars. Wing like the back, the white tips of the secondaries forming two bars across the wing. Primaries and tail, blackish brown. Bill, legs, and feet, like those of the male, perhaps slightly darker. Size, about the same.

Young.—Resembles the female, but the margins of the feathers are rusty brown, and the white wing bars are indistinct. Males have the sides of the head blackish.

Downy Young.—Crown of head, lores, and sides of face, dark brown; upper parts brown tinged with fulvous on upper part of back. Line over the eye and on each side of chin, white; the latter making a V-shaped mark. Under parts, pale brown, with center of breast and abdomen, white.

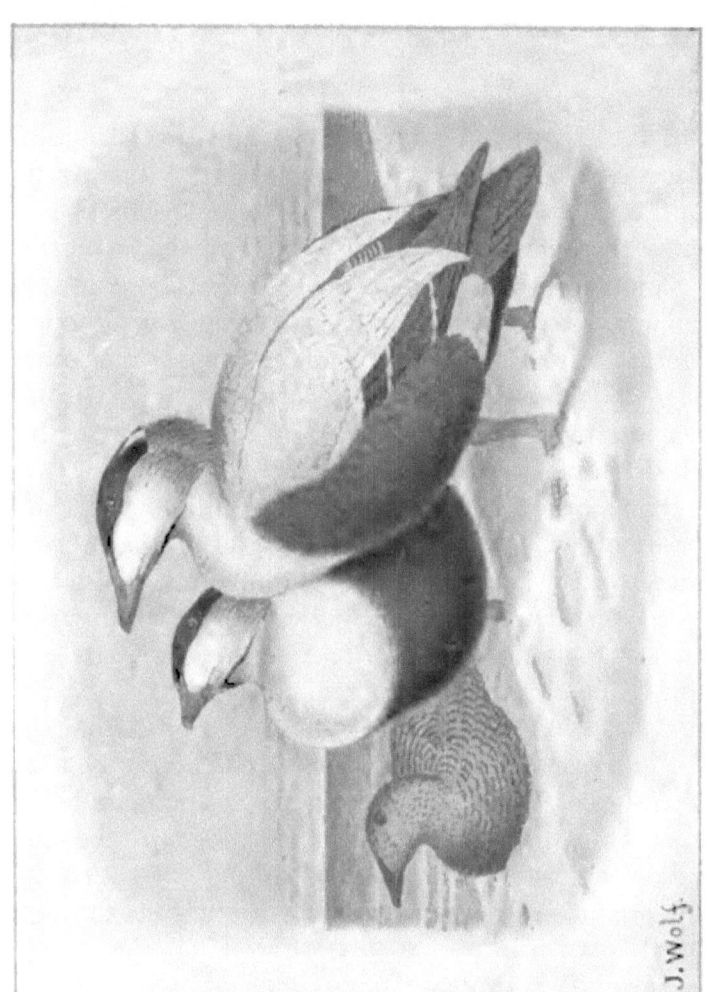

56. Pacific Eider.

PACIFIC EIDER.

FROM the Peninsula of Alaska, as far west as Attu of the Aleutian Islands, throughout the islands of Behring Sea, along the coast of Alaska to the Arctic Ocean, and eastward to the Coppermine River, is apparently the distribution of this species in the far northern region in which it finds its home. The principal breeding resorts are the islands of Behring Sea, although the birds nest also in great numbers in various other places. The habits are very similar to those of the Common and American Eiders, with the exception that this Duck does not breed in colonies.

Pretty much all that is known of the habits and economy of this fine bird is related by Dall, Nelson, and Turner, whose long residence in the bleak northwest afforded them ample opportunities for observing this Duck in its haunts. According to their reports the Pacific Eiders begin to approach the shores off the mouth of the Yukon River, if the ice permits, from the 10th to the 20th of May and proceed to choose the sites for their nests, the ponds and creeks in the marshes being at that time open. There does not seem to be any especial manifestation of affection during the courtship, all the preliminaries having probably been gone through with at sea, previous to the arrival of the mated birds near the shore. They come in small flocks, which break up into pairs, each couple resorting to the salt marshes. A moss-grown slope, a grassy tussock, or a depression made in the ground in some dry place near to a pond or creek, or even close to the sea-shore, is chosen as a site for the nest. The cavity is lined with grass or pieces of moss, and down from the parent's breast is provided in quantities, as the eggs are laid, sufficient to cover them, so that when the full complement is deposited the amount is very considerable. A peculiarity of this species seems to be that the eggs are not placed upon the down, but are merely covered by it. The male faithfully attends the female in the Arctic night until all the eggs are deposited, yet during the day he seems to lose all interest in matrimonial affairs, and joins other males which pass their time sunning themselves on the reefs near the shore. But in the twilight they fly silently back to their partners, to see if all is going on well. When upon the rocks the males keep uttering a long, guttural note which, when many are congregated together, sounds like a continuous grunting. The males outnumber the females, and several may be seen at times in attendance on one female. This is in Alaska, but near the mouth of the Anderson River, where MacFarlane found this species very abundant, the reverse was the case, and he was inclined to think they were polygamous, for he sometimes would find two females on the same nest. This, as has been already remarked, is occasionally witnessed among the breeding colonies of the eastern Eider.

When incubation has fairly begun the males retire to the sea and outlying rocks, and concern themselves no farther with their wives. The eggs are generally six or seven in number and are of a light olive drab color. Toward the end of June or beginning of July the young appear, and are at once led to the nearest water, generally a pond or creek, and later to the sheltered bays and mouths of rivers on the coast. The females now begin to moult, and like the young possess only one means of escape from their enemies—great skill in diving. The Eskimo amuse themselves at this time trying to strike the birds with spears, but are rarely able to hit one, so quickly do they vanish beneath the surface. The young are not able to fly much before the middle of September, and toward the end of this month all desert the main shores and are only found off the outer reefs and small islands. In the autumn it is said the male assumes a plumage very similar to that of the female, and the young males only attain the fully adult breeding dress at the commencement of the third year. As a rule the Pacific Eider is very shy and difficult of approach, except when on land during boisterous weather. At such times the birds gather on the rocks on the shore in large numbers, and the natives are accustomed to catch many by throwing hand nets over them. A bright night, when the wind is blowing hard, is the best for this purpose, and the flocks seem so stupid, as their members are all huddled together, that one is permitted to approach close to them. This species is also averse to flying in boisterous weather, and, as the body is heavy the birds appear to have difficulty in taking wing from the sea, and will flap along over the surface, and then all dive simultaneously. They descend to a great depth, and remain under water for a long time, swimming great distances before rising.

The principal food of this Duck is mussels and other bivalves, and it seeks these sometimes in water thirty or forty feet deep. During the breeding season, the note usually uttered when the sexes are together is a kind of *Coo*. The Pacific Eider is a handsome Duck, resembling somewhat the eastern species and weighs from four to six pounds. It seems to dislike stormy days and rough water, although it must get plenty of both during the year in the latitude it lives in, and at such times assembles in numbers along the beach, or on the rocks near the shore, or else swims about in the sheltered bays and inlets, where the force of the wind is not felt. The Eskimo name for the bird is Mi't hŭk. South of the mouth of the Yukon River the Pacific Eider plays a very important part, says Nelson, in some of the religious festivals of the natives, which occur in December. It is a kind of an Eskimo " harvest-home."

SOMATERIA V-NIGRUM.

Geographical Distribution.—Peninsula of Alaska, Aleutian Islands, and islands of Behring Sea, and along the Alaskan coast to the Arctic Ocean, and east to the Coppermine River. Breeding throughout its range.

Adult Male.—Plumage almost precisely like that of the two previous species, except that on the throat there is a very long V-shaped black mark, beginning on the chin and extending to a line intersecting the occiput. Very much longer and somewhat narrower than a similar mark on the King Eider. The black on the head is bordered beneath by pale sea green for nearly its entire margin, like that of the American Eider. The bill is different from that of the other Eiders, being broader and deeper through the base, while the frontal angles are much shorter and very acute. The extension of the feathers forward underneath the mandible (between the jaws) surpasses that on the sides, which is rather the reverse in the other species; but this can hardly be considered of specific value. The color of the bill is orange red on frontal angles and base of culmen, grading to orange toward the tip, which is yellowish white. Iris, dark brown. Legs and feet, brownish orange. Bill, from tip to end of frontal angle, $1\frac{8}{10}$ inches; greatest width of angle, $\frac{1}{2}$; culmen, $2\frac{1}{4}$. Total length, about 22 inches; wing, $11\frac{1}{2}$; culmen, $2\frac{1}{4}$.

Adult Female.—Head, chin, throat, and neck, pale brown, streaked with dusky; darkest on top of the head. Upper parts, rufous, barred with black, the bars broadest on back and scapulars; some of the latter and tertials tipped with yellowish white. Lesser wing coverts, dusky, tipped with white. Greater coverts, pale buff. Secondaries and primaries, blackish brown; the former having the edge of outer webs pale buff. Tail, blackish brown. Breast and sides, pale buff, barred with brownish black. Under parts, uniform grayish brown. Under tail coverts, barred with black and rufous. Wing, $11\frac{1}{4}$; culmen, $1\frac{3}{4}$; tarsus, $1\frac{3}{4}$.

Downy Young.—According to Stejneger, who obtained it on the Commander Islands, the downy young is precisely similar to that of the Common Eider, *S. mollissima*.

KING EIDER.

THIS Eider is a native of both the Old and New Worlds, and in North America is found across the continent in the Arctic regions, and comes south in winter on the Atlantic coast occasionally as far as New Jersey. It is not so abundant as any one of the other species of Eiders although large flocks are occasionally met with in the far northern regions. It is a boreal species and does not go very far south of its breeding places unless driven by stress of weather, when a few appear within the limits of the United States. It occurs at times on the Great Lakes in winter, and has been recorded from Illinois and Wisconsin, but does not frequent any part of the Pacific coast south of Alaska. In that Territory it is rare at St. Michael's, but is very common in Behring Straits, on the Siberian side, and near Waukareen and Tapkan and also on St. Lawrence Island. In the summer from Icy Cape on the Arctic Sea, and thence eastward, it occurs in large numbers, the birds being, however, chiefly males, as at that time the females are busy with their broods on the ponds and streams, away from the coast. It is the handsomest of the Eider Ducks, the delicate pearly gray crown of the head showing to great advantage with the other colors of sea green, black, and white of the head and neck, and deep buff of the breast, all contrasted with the bright orange of the bill. The nest is merely a depression in the ground near water, sometimes on the beach, and lined with down, on which are deposited usually six eggs, of a

57. King Eider.

light olive gray shade, sometimes grayish green. In its habits the bird does not differ materially from its relatives. The males desert the females when incubation commences, and assemble in great flocks by themselves upon the outlying reefs, or on the sea not far from shore, and are joined by the females and young in the autumn.

It seems, however, to be even more of a Sea Duck than the other Eiders, and is met with a long distance from land, on the open ocean. The males assume a dress similar to that of the females, after the breeding season, save that one or two pairs of white patches remain, by which the sex can be determined. The skin of this bird is used by the Eskimo for making clothing, and that of the female, split down the back and the head and wings removed, is placed inside the seal-skin boot and is very comfortable in winter. The King Eider feeds on fish and various kinds of mollusks, and as may be expected from such a diet, its flesh is not particularly palatable. In size it is somewhat less than all the other Eiders, except possibly the Spectacled or Fischer's Eider. As a diver, and possessing an ability to remain under water for a lengthened period, the present species is in no way inferior to its relatives, and a large portion of its time when at sea is engaged in exploring the depths, and seeking the marine creatures upon which it subsists.

SOMATERIA SPECTABILIS.

Geographical Distribution.—Northern parts of Northern Hemisphere. South on the Atlantic coast in winter to Georgia, and to the Great Lakes in the interior. Not found on Pacific coast south of Alaska. Breeding in the Arctic regions.

Adult Male.—A line along the base of the bill, and over and onto the anterior edge of the frontal process, a spot beneath the eye, an indistinct line at bottom of pearl gray on nape, and a broad V-shaped mark from chin along sides of the throat, jet black. Top of head and occiput, pearly gray. Yellowish white stripe over and behind the eye. Cheeks, pale sea green; this color extending on sides of head along the pearly gray until it fades away in white. Rest of head, chin, throat, neck, upper part of back, wing coverts (except the greater and outer webs of lesser coverts) and a large patch on each side of rump, white. Breast, dark cream buff, varying, however, among individuals in intensity. Greater wing coverts, scapulars, and primaries, brownish black; the scapulars and tertials sickle shape bending over the wing, and rufous along the shaft. Lower back, rump, upper tail coverts, and rest of under parts, black. Tail, brownish black. Bill varies considerably in shape at different periods of the year. In the breeding season, a high, square, soft process is elevated on the culmen between the base and the nostrils, and supported by some fatty substance. Matrimonial duties finished this shrinks, and the bill on its upper outline returns to nearly the normal Eider shape. On account of this protuberance the feathering on the maxilla is quite different from that of the other species of the genus, and on the elevated culmen nearly reaches the nostril, while on the side it extends but a short distance beyond the corner of the mouth. Bill and elevated process, reddish orange. Iris, yellow. Legs and feet, orange red. Total length, about 23 inches; wing, 11; culmen in front of process, $1\frac{1}{10}$; tarsus, $1\frac{3}{4}$.

Adult Female.—Head, chin, and throat, dark buff, streaked with dark brown, conspicuously on top of head and but faintly on the sides. Chest and sides, light buff, with irregular black bar on tip of feathers. Feathers of back and scapulars, blackish brown, with yellowish tips. Shoulder of wings, blackish brown; tips of feathers, rufous. Greater coverts and secondaries, black, with white tips, forming two narrow bars across the wing. Outer webs of tertials, rufous. Rump and upper tail coverts, dark buff, barred irregularly with black. Tail, black. Under parts, blackish brown; under tail coverts, rufous, with V-shaped black bars. Bill, greenish brown. Legs and feet, dull ochre. Total length, about 23 inches; wing, $10\frac{2}{3}$; culmen, $1\frac{1}{4}$; tarsus, $1\frac{3}{4}$.

Downy Young.—Resembles that of the Common Eider, but the upper parts are more rufous, and the cheeks, throat, and under parts more yellow.

58. Ruddy Duck.

RUDDY DUCK.

GENERALLY dispersed over all North America, the Ruddy Duck is found as far south as Cuba and Guatemala. It breeds throughout most of its range from Hudson Bay and Great Slave Lake in the north, and in the Mississippi Valley from Minnesota to Texas. Although it is common on many parts of the Pacific coast, it does not seem to go as far north on the west side as Alaska, and has never even straggled into the Eastern Hemisphere. It places its nest near some pond or other inland water, and constructs it of grass or dead leaves. The eggs are creamy white, and quite numerous, as many as twenty having been seen in a single nest, but this, it would seem, must have been the work of two females, as it is very doubtful if one alone could cover so many.

The Ruddy Duck is one of the sprightliest birds among our Water fowl, and at times presents a very comical appearance upon the water. It swims easily and rapidly, its enormous feet propelling the bird with considerable power. When on the water the body is deeply immersed, and if suspicious or alarmed, I have often seen it quietly sink beneath the surface without diving, and disappear. The Dabchick, or Hell Diver, has a similar way of vanishing. This species seems to have no preference for the quality of water it frequents, whether it is salt, fresh, or brackish. It usually goes in considerable flocks, and flies with great swiftness, turning first the upper side of the body, then the under, to the spectator as it rushes along. It is very erratic in its ways, and exceedingly quick in its movements, whether on the water or in the air. It walks fairly well, and takes wing from the land at once, but has considerable difficulty in rising from the water, and is obliged to run along the surface, beating it with both feet and wings, before it can get away. It is a most expert diver and is able to stay long, and go far under water. When swimming it has the habit of elevating its short, stiff, spiny-looking little tail straight up in the air, sometimes inclining it forward toward the head, and as the latter is very large as is also the bill, and is held well back, there seems hardly enough body between them to sustain all this superstructure, especially as the bird swims so deeply that a large portion is hidden beneath the surface. In this position the male, for he is the one that exhibits himself usually in this way, moves up and down among the others as if challenging their admiration. It is a very gentle species, and plunges into the decoys with a slide and a splash like the Buffle Head or Hooded Merganser, or other of the small rapid-flying Duck.

When in flight this species makes a whirring sound caused by the rapid movements of its concave wings, as it buzzes along, the members of a flock twisting and twirling about, but going usually in a straight line, and they seem more like a swarm of bees than a bunch of Ducks. Their flight is so swift, and the body is comparatively so small, that they are by no means an easy bird to shoot, and much allowance must be made for the rapidity with which they hurl themselves through the air. Formerly but little attention was paid to this Duck by sportsmen; it was so small that it was allowed to go by unheeded; but of late years, on account of the growing scarcity of larger Ducks, it is more sought after, and many are killed. It is accustomed to frequent bays and coves, and a number of gunners will assemble in boats and line the entrance to one of these, and gradually advancing, close in upon the birds, which are shot either on the water while swimming or as they attempt to fly past. There is nothing sportsman-like in this proceeding, but as the birds bring now quite a sum per pair in the market, it is killing merely for gain. At times, when one of these battues was going on in Currituck Sound, it seemed as if the country was being bombarded by a hostile fleet, so frequent and heavy were the explosions. Of course if this practice is continued, it will have one of two natural consequences: either the extermination of the species in that locality, or its removal to more secure situations.

The male Ruddy Duck in full summer dress is a very handsome bird, and resembles very little the same individual in the costume he usually wears in winter. The black head and nape, with the large white patch upon the face, are well contrasted with the rich dark red of the upper parts of the body and the silver grayish white of the lower plumage. When so arrayed he is an object of singular beauty, but unfortunately he only exhibits himself in these nuptial garments for a brief period in the year.

This species has a very great number of names, one apparently for almost every locality it visits. Some of these are, Broad Bill Dipper, Coot, Broad Bill Coot, Bumble Bee Coot, Heavy-tailed Duck, Salt-water Teal, Booby, Booby Coot, Stiff Tail, Spine Tail, Ruddy Diver, Ruddy, Stick Tail, Bristle Tail, Bull Neck, Steel Head, Rook, Greaser, etc. Of late this Duck has become quite fashionable among the gourmets of the cities, and is considered apparently as desirable as some of the larger

Ducks of extended reputation. This idea has been taken advantage of by the market men, and a pair of this small inferior Duck bring as high a price as Red Heads did a few years ago. While affording a fair dish, if properly broiled, there is nothing in the flesh of this bird to merit any particular commendation. Its food usually consists of various grasses, roots, and leaves of plants, and possibly at times it may vary its diet with mollusca of different kinds. Being a diving Duck, it obtains the articles for its bill of fare from off the bottom.

ERISMATURA JAMAICENSIS.

Geographical Distribution.—North America generally, except Alaska. South to the West Indies and Colombia. Breeds throughout the greater part of its range, from Hudson Bay to Guatemala.

Adult Male in Full Plumage.—Upper part of head, including the eye and nape, glossy jet black. Sides of head and chin, white. Throat and all the neck, back, upper tail coverts, scapulars, and flanks, bright reddish chestnut. Wing coverts, lower back and rump, grayish brown. Primaries, dull brown, speckled near edge of outer web with gray. Tail, brownish black. Under parts below the upper part of breast, silvery white, which is the hue of the tips of the feathers only, the hidden portion being brownish gray. Sometimes these tips wear away, and then the under surface appears mottled. The breast is tinged with rust color; this in some specimens appearing also on the abdomen. Under tail coverts, white. Bill and eyelids, grayish blue. Iris, hazel. Legs and feet, grayish blue; webs, dusky. Total length, about 16 inches; wing, 6; culmen, $1\frac{8}{10}$; tarsus, $1\frac{3}{10}$.

Adult Female.—Upper half of head, including the eyes, dark brown; in some individuals there are blackish feathers, tipped with reddish chestnut. Cheeks, brown, but lighter than top of head. A white stripe from below the eye, sometimes distinguishable almost to base of bill, goes to the nape. Chin, white. Throat and neck, brownish gray; tips of feathers on lower neck in front, white. Upper parts, dusky brown, mottled and speckled with grayish buff. Lower parts, silvery white, this hue produced as in the male by the tips of the feathers. A yellowish wash on upper part of breast. Sides and flanks, barred with brown. Wings and scapulars, dark brown, the latter speckled with paler brown. Tail, dark brown; in some individuals the feathers are broadly margined with pale olive or grayish brown. Under tail coverts, white. Bill, blue. Legs and feet, bluish gray; webs, dark. Total length, $15\frac{1}{2}$ inches; wing, $5\frac{1}{2}$; culmen, $1\frac{6}{10}$; tarsus, $1\frac{1}{4}$.

Young Male.—Only differs from the adult female in having the sides of the face more or less white, sometimes entirely so, and sometimes the white is spotted with brown and black.

Young.—Has top of the head like that of the female; sides of the head, dark brown, with a white stripe from base of bill, where it is broadest, to the nape, passing below the eye. Chin and throat, whitish. Neck, brownish white, many downy feathers protruding among the full grown. Back and scapulars, blackish brown, barred with reddish buff. Middle of the back and rump, reddish brown. Upper tail coverts, blackish brown, barred with reddish, like the scapulars. Sides and flanks, with the tips of the feathers, yellowish; other parts, dusky. Under parts, silvery gray, passing into dusky, on the crissum. Under tail coverts, white. Maxilla, dusky; mandible, yellow.

Downy Young.—Head and upper parts, smoky brown, darkest on head; a brownish white stripe from bill to occiput below the eye, bordered beneath by one of dusky brown. Breast, sooty brown; under parts, grayish white.

MASKED DUCK.

THIS rather handsome Duck is a native of the West Indies and South America, and it is only as an accidental visitor within the limits of the United States that it can be included in our Fauna. A few instances only are on record of its capture within our borders; at Lake Champlain in New York, Malden in Massachusetts, and on Rock River, Wisconsin, widely separated localities. Another individual was supposed to have been seen on Lake Koshkonong, Wisconsin, but as it was not secured its identification was impossible. It has also been procured near Brownsville in Texas, and at Matamoras in Mexico. In Trinidad and the northern parts of South America, in some of the West Indian Islands and in Western Mexico this Duck is frequently met with and in some localities is not rare. Nowhere, however, is it observed in such large flocks as are frequently seen of its relative, our common Ruddy Duck, in Northern waters.

Like that species the flight of the Masked Duck is rapid, but not sustained for any great distance, and it is a sociable species and loves to keep together in small companies on the lakes and lagoons in the localities it inhabits. In Trinidad its flesh is considered excellent, and it is regarded with much favor. As a diver it is an expert, and remains under water for a long time. It swims deeply like the Ruddy Duck, but on land is awkward, usually holding itself upright and supported in a great measure by its stiff tail. It is a handsome bird with a

59. Masked Duck.

more striking plumage even than the summer dress of
the Ruddy Duck. Occasionally this species straggles far
to the southward in South America, and it has been procured in the Argentine Republic and in Chili, but this
must be regarded in the same light as its various appear-
ances in our northern waters, merely as instances of a few
individuals having strayed, from exceptional causes, far
away from their usual habitats.

NOMONYX DOMINICUS.

Geographical Distribution.—Tropical America, from the West
Indies and northern South America to the Lower Rio Grande;
straggling occasionally as far north as Wisconsin and Massachusetts, and south to the Argentine Republic and Chili.

Adult Male.—Head, excepting nape, and chin, intense black.
Nape, throat, neck, back, scapulars, and upper tail coverts, dark
rusty cinnamon; center of feathers, black, showing conspicuously.
Lower back and rump, dark brown spotted with black, and some
feathers edged with white. Upper part of breast, uniform, dark
rusty cinnamon grading into pale reddish buff. Sides and flanks
darker, with black centers to the feathers. Wings, blackish
brown, with a long, narrow, white speculum. Under tail coverts,
cinnamon blotched with black. Tail, dark rufous brown; shafts
of feathers, black. Bill and eyelids, pale blue; median line on
maxilla, nail, and bare skin of chin, black. Mandible, reddish
white; tip, black. "Outer aspect of tarsus and two outer
toes, dark brown or black; the inner side of the tarsus, inner toe,
and membranes, pale brown spotted with black" (Gundlach).
Iris, dark brown. Total length, about 15 inches; wing, $5\frac{1}{4}$;
culmen, $1\frac{8}{10}$; tarsus, 1. *Description taken from individual
killed at Malden, Mass., in 1889, and now in the Field Col-
umbian Museum, Chicago.*

Adult Female.—Top of head, stripe from base of bill through
eye to occiput, and one from gape to occiput, black. Superciliary stripe and rest of head, buff, becoming whitish on chin
and throat. Neck, buff mottled with brown. Upper parts,
black, feathers edged with deep buff. Wings, dark brown,
feathers tipped with yellowish white. Speculum, white. Pri-

maries and tail, brownish black. Under parts, ochraceous spotted with blackish on breast, flanks, and anal region. Abdomen, uniform ochraceous. Bill, horn brown; nail, black. Total length, about 13 inches; wing, 5; culmen, $1\frac{8}{10}$; tarsus, 1.

Young Male.—Sides of head, mottled with buff, and the under parts of the body are whitish. In other respects the specimen agrees with the adult male. Still younger birds resemble the female, but the feathers have no brown centers on breast and sides, and the under parts are paler generally.

60. American Merganser.

AMERICAN MERGANSER.

KNOWN by its various names of Goosander, Buff-breasted Sheldrake, Buff-breasted Merganser, Swamp Sheldrake, Weaser, Fish Duck, American Merganser, Scie de Mer and Sea Sawbill in Louisiana, and many others in various parts of the land, the present species is distributed throughout the whole of North America, breeding in the West as far south as Northern Colorado, and occasionally going to Alaska and certain of the Aleutian Islands. It has also visited the Bermudas. In Alaska it is only known to have occurred a few times within the Territory, but it appears to be an accidental visitor at Unalaska Island.

This Merganser is the largest, and in my opinion the handsomest of the Saw-bill Ducks, so-called from the curiously lengthened bill lined on the edges with serrations like the teeth of a saw. The American Merganser resembles almost precisely the European species, and it is very doubtful if anything is gained scientifically or otherwise, by the attempt to separate them; the difference being that the European bird has an *exposed*, the American, a *concealed*, black bar across the wings. The Goosander breeds in the hollows of trees, except in far northern districts such as certain portions of the Arctic regions where trees sufficiently large are scarce, and there it makes its nest upon the ground. Generally a large tree is selected upon the borders of some inclosed lake among the mountains, or on the bank of a river in a lonely, retired situation, and in a hollow, perhaps twenty feet from the ground, the eggs are deposited and the young hatched. For so large and heavy a bird, it is very quick and agile, and I have seen it dart among the trees, and enter and leave the nest with an easy dexterity that was surprising. It alights and walks upon the branches without any difficulty, and it is a curious sight to observe so large a member of the Duck tribe living upon the trees.

This bird is the rarest, I think, of all the Mergansers or Fish Ducks. At all events that is the case in the Atlantic States, but it is much more frequently met with in different parts of the West and on the Pacific coast. When it appears in the autumn coming from its northern breeding grounds, it arrives in flocks of considerable size, but in a short time these break up into small parties, and keep by themselves, for in my experience, the Goosander does not often associate with other Ducks, but seems best satisfied with the company of two or three of its own species. It flies with great rapidity usually in a direct line, if over water, but if in the woods, twists and turns among the trees and dodges the intervening branches with the dexterity of a wild pigeon. The eggs are buffy white, and the young are carried down to the water by the mother in her bill. The little things are most expert swimmers and divers from the moment they enter the water, and require no teaching to become proficient in these accomplishments. They follow the mother closely, either huddled around her in a compact mass, or strung out behind her, snatching insects from the surface of the water. If alarmed they scurry away with a speed that is marvelous, running in fact over the bosom of the lake or river; the flock leaving a wake behind them like that of a miniature boat. They do not dive unless hard pressed, but trust at first to skimming surface of the water before their pursuers, and generally easily outstrip a boat, leaving it far behind. But if cornered in any way, by being forced into a narrow bay, or brought close to the shore, they will then dive and remain out of sight for a considerable period, coming in view again long distances from where they disappeared. It is no uncommon sight to witness the female swimming quietly along with most of her family snugly and comfortably settled upon her back. The little ones becoming tired, the mother sinks her body until her back is on a level with the surface, when the young swim or clamber on to it, and she rises, lifting them out of the water. Occasionally the whole family will settle themselves upon a sand bar in the middle of the river or lake, or on a gravelly beach near the bank, and preen their feathers and sun themselves; but at the least alarm they immediately take to the water and move rapidly away, for in disposition they are very wild birds.

The food is exclusively fish, which are pursued and seized under water, and immense numbers are destroyed by this species. The Goosander is tenacious of life, and requires large shot to bring it down, and frequently, after falling, it recovers itself and effects its escape. It is a handsome bird, and in life the under part of the body is suffused with an exquisite roseate tinge or glow, that fades rapidly after death.

MERGANSER AMERICANUS.

Geographical Distribution.—Throughout North America, breeding in the United States, and in the northwest.

Male.—Head and neck, shining blackish green, crest on occiput. Upper parts, black; rump and upper tail coverts, ash gray. Primaries and secondaries, black; rest of wing, mostly white, with a black bar crossing it, formed by the bases of the greater coverts. Under parts, rosy salmon color, which fades rapidly after death. Tail, ash gray. Bill and feet, vermilion; the hook, black. Iris, carmine. Length about 26 inches; wing, 10.75; tarsus, 1.95; culmen, 1.95.

Female.—Head and neck, reddish brown; an occipital crest of lengthened feathers of the same color extends along hind neck; chin and throat, white. Upper parts, ash gray. Primaries, black; terminal half of secondaries, white, forming a speculum or spot on the wing. Flanks, ash. Lower parts, pale salmon color in life; white in preserved skin. Tail, ash gray. Bill, red; culmen, blackish. Feet, orange; webs, dusky. Iris, yellow. Average total length, $22\frac{1}{2}$ inches; wing, $9\frac{8}{10}$; tarsus, $1\frac{5}{10}$; culmen, $1\frac{9}{10}$.

Distance from nostril to nearest feather on head GREATER *than height of the maxilla at base, in both sexes.*

Downy Young.—Upper parts, hair brown, with four white spots. Half of head above and hind neck, rusty. Upper part of lores crossed by a brown stripe, and a white one on lower part, bordered beneath by a narrow one of brown. Rest of head and neck and entire under parts, white.

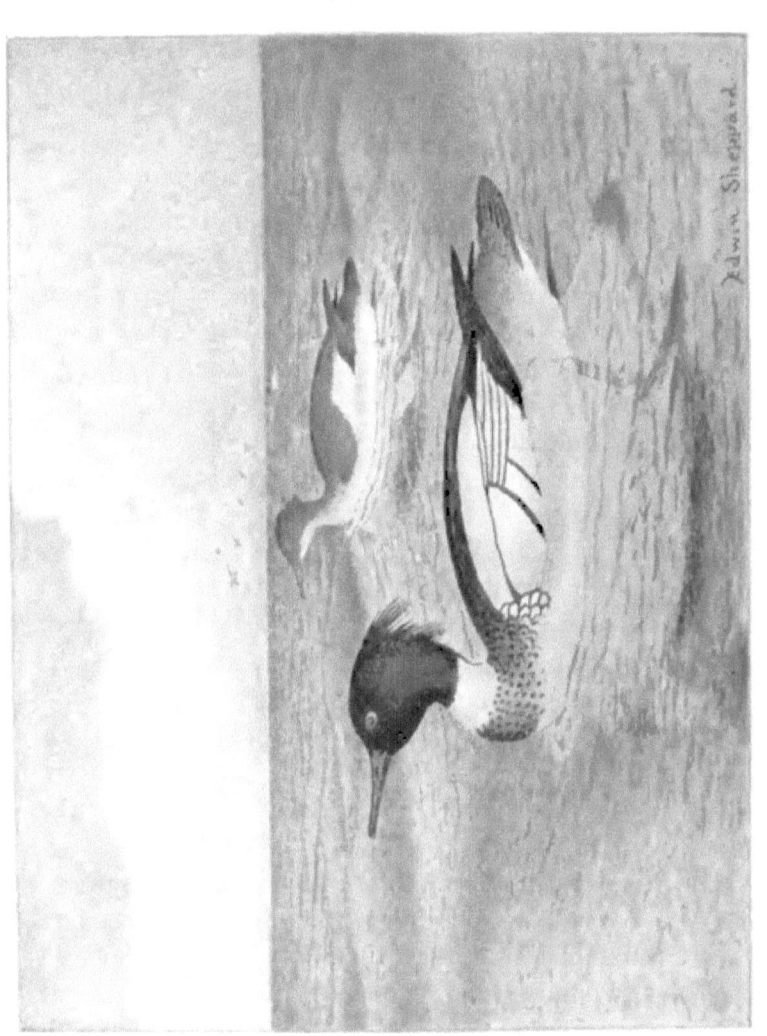

61. Red-Breasted Merganser.

RED-BREASTED MERGANSER.

THIS well-known species is an inhabitant of both the Old and New Worlds, and the birds of the different hemispheres, unlike the Goosander, have been permitted to remain as one species, not even the slightest character having been discovered whereby they could be separated. It is known in many parts of our country by various names, those most commonly employed perhaps being, Sheldrake, Fisherman, Fish Duck, Shelduck, Saw Bill, Pied Sheldrake, Big Hairy Crown, Red-headed Merganser, and the one at the head of this article. In North America it breeds from the Northern States in the Union, as far as the Aleutian Islands and coast of Alaska, and is common in the district of St. Michael. Mr. Turner found it abundant in the islands of Attu and Atkha of the Aleutian chain. It arrives there in the latter part of May or beginning of June, and remains through the summer; and the young are fully fledged in September. The Aleuts consider its flesh a great delicacy, and it is more highly prized by them than any other Duck. In winter it migrates as far as southern California on the Pacific coast, and to Florida on the Atlantic.

This Merganser is more of a marine species than the Goosander, and is frequently met with on our coasts, and up the rivers that empty into the sea. Its nest is placed upon the ground, generally hidden under a bank, or some rock or fallen trunk of a tree, and is formed of grass, together with feathers and down plucked from the parents' breast. The eggs, usually seven to ten in number, are a fawn, or bright cream color. The young are very active, follow the female on the water, and scurry away at the least alarm in the manner already described of the young of the Goosander. The Red-breasted Merganser flies with great rapidity and makes very little noise with its wings, and I have had it approach when I was in a blind, so quietly that its appearance, in front of me and close to the bank, would be the first intimation given that any were in the vicinity. When startled or alarmed, either while flying or swimming, they are in the habit of uttering several low, guttural croaks, resembling in no way the quack of a Duck, and, if on the water, they dive quickly and sometimes remain beneath the surface for a long time, appearing in quite a different place from that expected. They seem to be very observant, and frequently I have noticed a small flock, passing rapidly along the shore, suddenly turn and retrace their way and alight with a splash, and immediately dive and commence to feed. It would seem that the birds must have seen a school of small fish as they flew by, and returned to take advantage of their presence.

When swimming along both sexes are accustomed to elevate and depress the long occipital crest, giving them alternately a trustful and wild appearance. This species feeds entirely on fish, and the flesh consequently is rank and of a very disagreeable flavor. When engaged in fishing, by their rapid diving and maneuvering beneath the waters, they cause the small fish—if the schools are of any size—to become widely scattered, and many rise close to the surface. The Gulls take advantage of such opportunities, and pounce upon their luckless finny prey from above, and then, with Ducks diving into the depths and Gulls plunging from above, the scene is a very lively one. I remember on one occasion watching a number of this Merganser engaged in fishing in a cove, when their movements attracted to them a large flock of Bonaparte's Gull (*Larus philadelphia*), which hovered over the Ducks for a moment and then began to plunge head foremost into the water, one after another in rapid succession, emerging frequently with a small fish in the bill. The Mergansers paid no attention to their fellow-fishermen, although at times a plunging Gull would come perilously near one of the saw-billed gentry as he rose from the depths; and what with the rising and disappearing Mergansers, and the air above them filled with the forms of the darting Gulls, executing all manner of swift and graceful evolutions, the scene was very spirited and full of animation. Although having a great partiality for the sea-coast, and the bays and rivers adjacent to the ocean, this Merganser is also found, perhaps in not so large numbers, in the interior of the United States; and among certain of the Wisconsin lakes is of regular occurrence, as it passes north and south on its annual migration in the spring and autumn. The males generally precede the females, each sex traveling toward their breeding grounds apart from the other. The female of this species and that of the Goosander are very much alike in the general color of their plumage, and one might readily be mistaken for the other; but the Key indicates how each can be distinguished. The female of the Goosander, however, is a little the larger.

The Red-breasted Merganser is not uncommon in many parts of the British Islands and on the continent of Europe. It is also found in Greenland and Iceland, and goes eastward as far as Formosa, China, and Japan; in fact, has a fairly general distribution over the northern parts of both hemispheres. It is one of the Duck tribe most frequently met with by the sportsman, especially on the sea-coast, when engaged in his favorite pastime of shooting over decoys, and while prized by some, is not considered by many as an especially desirable addition to the game bag. The male, however, is one of the handsomest Ducks in our country, and with his glossy metallic head and crest, and variegated body, presents a very brave appearance as he swims proudly along by his mate under the bright sun of the early spring.

MERGANSER SERRATOR.

Geographical Distribution.—Northern portions of both hemispheres. In winter throughout the United States. Breeds from the northern States to the Aleutian Islands.

Male.—Head and occipital crest of lengthened hair-like feathers, black, with green and purple reflections, the former predominating. A broad white ring around the neck beneath the black, with a narrow black line crossing it at back. Back and inner scapulars, black. Lower back and rump, gray, mottled with black and white. Primaries, blackish brown. Wing, mostly white, crossed by two black bars, formed by the bases of the secondaries and greater coverts. Outer webs of inner secondaries, edged with black. In front of the shoulder of the wing is a patch of white feathers narrowly bordered with black. Lower neck and upper part of breast, pale cinnamon, or dark brownish buff, streaked with black. This conspicuous band varies in depth of coloration among individuals. Flanks, irregularly barred with narrow lines of grayish white and black. Rest of under surface, white, suffused with a salmon tinge. Tail, grayish brown, lighter on edges of webs. Bill, carmine, with the culmen dusky; nail, yellowish. Legs and feet, orange red. Iris, carmine. Average total length, about $22\frac{1}{2}$ inches; wing, $8\frac{1}{10}$; tail, 4; tarsus, $1\frac{8}{10}$; culmen, $2\frac{1}{10}$. *Distance from nostril to nearest feather on head* LESS *than height of bill at base, in both sexes.*

Female.—Top of head and crest, fuscous; sides of head and neck, brownish buff or pale cinnamon. Upper parts, dark grayish, inclining to a brownish hue. White patch on the wing, divided by a black bar formed by the bases of the secondaries. Throat, white; lower neck, gray. Under parts, white, tinged with salmon. Bill, legs, and feet, similar in color to those of the male, but less bright in hue. Length, about 20 inches; wing, $8\frac{1}{2}$; tarsus, $1\frac{1}{2}$; culmen, $2\frac{2}{10}$.

Young.—Chin and throat, pale reddish; lower neck and upper part of breast, brownish white. Base of secondaries, black, forming bar across the wing. Rest of plumage, similar to that of the female.

Downy Young.—Sides of head and neck, cinnamon, inclining to rusty, becoming lighter on the lores, which are bordered above and below with a dusky stripe. Upper parts, hair brown; cheeks, spot on wing, and on each side of back and rump, and also all the lower parts, yellowish white.

HOODED MERGANSER.

WATER Pheasant, Hairy Head, Hairy Crown, Swamp and Pond Sheldrake, Cock Robin, Little Saw Bill, Saw-bill Diver, Spike Bill, Wood Duck, Bec Scie and Cotton Head in Louisiana, and Hooded Merganser are some of the names by which this beautiful bird is known to the gunners and sportsmen of the United States. Numerous others are also given it, some of which are extremely local, and never heard save by a very few. It is much smaller than the two preceding species of Merganser, and the male is remarkable for the large and beautiful crest, white, margined with black. It is exclusively a North American species, and has only appeared at rare intervals in the Old World, where it can be regarded merely as a straggler. It ranges all over North America from Alaska and possibly Greenland, on the respective sides of the continent, to Mexico and Cuba. In Alaska it is rare and probably only wanders up to that Territory in the summer time in small scattering flocks, but is very common in the United States, breeding in many parts of the land, even as far south as Florida, and spreading all over the Union in autumn and winter. This species, like the Goosander, breeds in hollow trees, lining the cavity with grass, dry leaves, and feathers, and down from the female's breast, and about six ivory white eggs are deposited. The site for the nest is generally in some tree standing on the border of an inland lake or stream in the forest, where discovery would be least likely, and

62. Hooded Merganser.

where small fish in the near-by waters would be most abundant. They consume immense numbers of fish, and the presence of a few Mergansers, no matter of what species, on a trout lake or stream, means great loss to the sportsman, as the fry have no chance of escaping the rapid movements of these hungry, energetic birds.

Their progress under water is extremely rapid, and the wings as well as the feet are used as means of propulsion, perhaps more dependence being placed upon the wings, and they may be said to fly beneath the surface. The female carries the young down to the water in her bill, and the little creatures are at once entirely at home in the element; diving, and sporting with each other as if they had become perfected by long practice, instead of its being their first experience.

The Hooded Merganser appears to be equally as numerous in the autumn and winter in the interior of the United States as on the sea-coast, and frequents the lakes in company with the larger species of Ducks, or is seen rapidly passing over the surface of the rivers. On the wing it is one of the swiftest Ducks that fly, and it hurls itself through the air with almost the velocity of a bullet. Generally it proceeds in a direct line, but if it is alarmed at any object suddenly appearing before it, the course is changed with the swiftness of thought, and a detour made before again taking the first line of progression. Sometimes, without apparent reason, the course will be altered, and away it shoots at right angles to the first route; and again, it vacillates as though uncertain which way to take, or as if it was looking for a good feeding place. Usually five or six, but more frequently a pair, are seen flying together, and often, on dull days when the lookout in a blind is somewhat relaxed, and the sportsman is consoling himself for lack of birds with possibly a nap or the lunch basket, the first intimation of the presence of a Hairy Crown is given by one or more flashing close over head with a startling whirr, and then as rapidly disappearing in the distance. It requires a steady hand and correct eye to kill them on the wing, and the gunner must be ever mindful of the good old adage in duck-shooting, " Hold well ahead!"

The movements of this bird upon the water are quick and active, and it swims rapidly and dives with great celerity. It is a beautiful object, and few birds surpass the male in attractiveness as he swims lightly along, elevating and depressing his beautiful crest. If suspicious, this species will sink the body until the water is almost level with the back, and sometimes disappears beneath the surface, apparently without effort, as if some unseen hand was pulling it down. When wounded it is one of the most difficult birds to secure, and it dives with such quickness, remains under water so long, and skulks and hides with so much skill that it is very apt to make its escape, and always tries the patience of its pursuer, whether dog or man, to the utmost. On the sea-coast the Hooded Merganser keeps mainly to the creeks and ponds in the marshes, and rarely is seen in the more open waters of the sounds, unless obliged to fly over the broad expanse when passing from one marsh to another; but it rarely alights far from any shore. It is fond of pursuing its finny prey under the shelter of a bank, or in quiet stretches of narrow, sinuous creeks, where it is least likely to be observed. It rises from the water without any preliminary motions, and is on the wing at once, and in full flight, the pinions moving with a rapidity that almost creates a blur on either side of the body, the outline of the wing disappearing. It utters a hoarse croak, like a small edition of the note of the Red-breasted Merganser. Altogether this handsome species is a sprightly, attractive creature, and a great ornament to the localities it frequents.

LOPHODYTES CUCULLATUS.

Geographical Distribution.—Throughout North America, from Alaska, and possibly Greenland, to Mexico and Cuba. Accidental in the British Islands, and the Continent of Europe. Breeding throughout its range.

Adult Male.—Head, neck, and back, black; crest, pure white, bordered narrowly with black. Scapulars, black. Wing coverts, dark gray, white patch on wing divided by a black bar. Tertials, black, with a white central stripe. Primaries, dark brown. Rump, dark brown. In front of wing on the side of chest are two black and two white crescentic bars, pointed at one end; the first on chest, the latter on back. Flanks, grayish brown toward the chest, grading into reddish brown toward the tail, crossed by fine wavy black lines. Under parts, pure white. Vent and under tail coverts, mottled with dusky. Bill, black. Legs and feet, yellowish brown. Iris, bright yellow. Total length, about 18 inches; wing, $7\frac{1}{2}$; tail, $4\frac{8}{10}$; tarsus, $1\frac{1}{10}$; culmen, $1\frac{1}{2}$.

Adult Female.—Head, neck, and upper parts, grayish brown, darkest on the back. Crest, reddish brown. Chin and throat, white. Patch on wing, white, crossed with a black bar. Flanks, grayish brown. Under parts, white, crissum with rather indistinct grayish brown bars. Tail, dark grayish brown, like the back. Bill: maxilla, black, edged with orange; mandible, orange; nail, brownish black. Feet, light brown. Iris, hazel. Length, about $16\frac{1}{4}$; wing, $7\frac{8}{10}$; tarsus, $1\frac{8}{10}$; culmen, $1\frac{1}{2}$.

Immature Male.—Head and neck, grayish brown, the latter mottled and blotched with black. Crest, brownish white, edged with blackish brown. Upper parts, blackish brown, all the feathers tipped with pale brown. Wings, colored like the back, a few of the tertials having a white stripe in the center, and the outer webs changing to black. Rump and upper tail coverts, dark umber brown. Primaries, blackish brown, the webs edged with pale brown. Breast, light brownish gray. Flanks, light brown. Lower breast, abdomen, and vent, white. Under tail coverts, blackish brown. Tail, dark brown, feathers edged at tip with brownish white. The feathers have a glossy appearance, but only give a slight indication of the plumage assumed by the adult male.

Downy Young.—Upper parts, brown, darkest on back and rump; lower portion of head, chin, and throat, light buff. Grayish white spot on either side of back and rump. Breast, pale brown; belly, white.

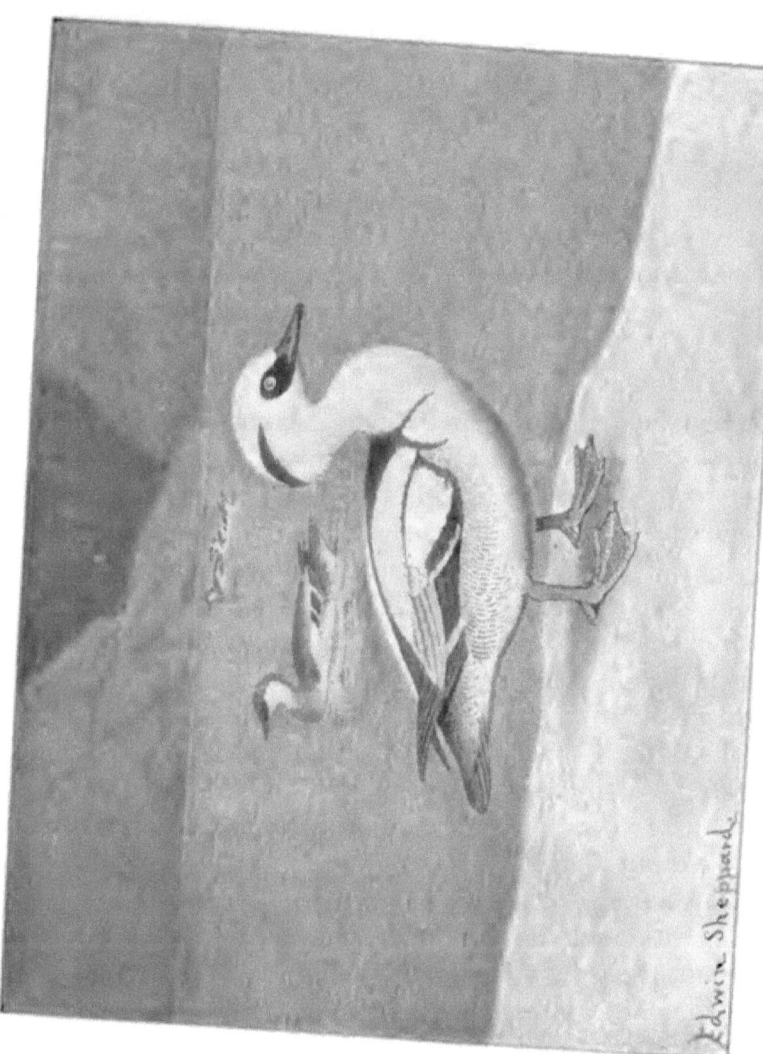

63. Smew.

SMEW.

IT is, so to speak, rather stretching a point, to include this beautiful species among the North American Water Fowl, with only an example of a female in the British Museum, purchased from the Hudson Bay Company, to prove the propriety of such a course. But I have always observed that ornithological committees are most lenient when the admission of a handsome bird (which under the most favorable circumstance can be regarding as the merest exceptional straggler from foreign lands) into their native avi-fauna is to be considered. I must, however, warn my American readers not to go hunting after this bird, for it is more than doubtful if any one of them will ever see it in the flesh within the limits of North America, unless shipped there from some port in the Old World. It is true that Audubon claimed to have obtained a specimen, and this also a female, on Lake Barataria in Louisiana near New Orleans in 1817, but none has been observed within the limits of the United States since that date so far as I am aware. At all events one cannot fail to notice that, up to this time, the male has rigorously and successfully avoided our shores.

The Smew is a native of northern Europe and Asia, going in winter to the Mediterranean, and from Great Britain on the west to Japan in the east. It is fond of resorting to fresh water, and frequents rivers and lakes, flies with great rapidity, and like all of its kind is a great diver. It feeds on small fish, shell fish, small reptiles, and insects. The Smew breeds in holes of trees, near lakes or rivers, retiring from the sea-coast during the mating season. The male is a very attractive bird, and in spite of the more brilliant coloring possessed by its relatives, in its pure white dress with the jet black markings, has a strong claim to be considered as one of the handsomest of them all.

MERGUS ALBELLUS.

Geographical Distribution.—Northern Europe and Asia, going in winter to the Mediterranean, northern India, China, and Japan. Very accidental in North America, the male never having been seen within its limits.

Adult Male.—General plumage, white. A large patch at base of the bill, including the lores and eyes; lower portion of nuchal crest, middle of the back, and two crescentic narrow lines on side of breast, outer edge of scapulars, and rump, jet black. Upper tail coverts, gray; edges, lighter. Middle wing coverts, white; greater coverts and secondaries, black, tipped with white. Primaries, blackish brown. Tail, dark gray. Sides and flanks undulated with fine black lines on a gray ground. Bill, bluish; nail, lighter. Iris, bluish white. Legs and feet, bluish lead color; webs, darker. Total length, about $16\frac{3}{4}$ inches; wing, $7\frac{6}{10}$; culmen, $1\frac{1}{4}$; tarsus, $1\frac{1}{2}$.

Adult Female.—Head and nape, chestnut brown; lores and cheeks, brownish black. Throat and sides of neck, white. Upper parts, brownish gray, darkest on the rump; some feathers on back tipped with ashy gray. Wings like the male. Tertials, brown. Upper breast, slaty gray. Sides and flanks, brownish gray. Under parts, white. Tail, brown gray.

Downy Young.—Upper half of head, including the eye, back of neck, and upper parts of the body, blackish brown. Sides of head, chin, and throat, small spot below the eye, a spot on edge and another at joint of wing, one on flanks and one on each side of the rump, together with the breast and abdomen, white. Upper breast, dusky; flanks, brown.

APPENDIX.

KEYS TO THE SUBFAMILIES, GENERA, AND SPECIES.

ORDER ANSERES.

FAMILY ANATIDÆ.

Bill usually flat, broad, sometimes long and narrow, provided with lamellæ, or tooth-like projections on sides and with a nail at the tip. Toes, four; the three anterior ones webbed, hind toe normal or lobed. Tarsus, flattened.

KEY TO THE SUBFAMILIES.
(For North American Species.)

A. Bill not compressed; no tooth-like serrations.
  a. Hind toe not lobed.
    a'. Neck very long, sometimes as long as the body. Size large. Skin covering bill extending to the eyes. — THE SWAN. *Cygninæ*.
    b'. Neck moderate. Bill strong, higher at base than at side; cutting edges more or less beveled, sometimes exposing the prominent lamellæ. — THE GEESE. *Anserinæ*.
    c'. Neck, short.
      a''. Tail feathers long, broad, rounded at tip. Upper coverts very long, nearly reaching end of tail. Bill rather narrow, high at base, tapering to a point. — WOOD DUCK. *Plectropterinæ*.
      b''. Tail feathers moderate, median pair sometimes elongated. Bill flat, frequently very broad. — FRESH-WATER DUCKS. *Anatinæ*.
  b. Hind toe broadly lobed.
    a'. Tail feathers normal. — SEA-DUCKS. *Fuligulinæ*.
    b'. Tail feathers narrow, stiff, pointed. — SPINE-TAIL DUCKS. *Erismaturinæ*.

B. Bill greatly compressed; serrations, tooth-like. — MERGANSERS. *Merginæ*.

FAMILY ANATIDÆ.

SWAN, GEESE, DUCKS, AND MERGANSERS.

THIS great family, represented throughout the world, contains the Swan, Geese, and Ducks, including the Mergansers or Saw-billed Ducks so-called. At one time North America was inhabited by myriads of these fowl, which passed throughout the length and breadth of the continent during spring and autumn in countless numbers; but of late years their ranks have been greatly thinned, and it is evident to the most casual observer that the birds are rapidly passing away. Of the many subfamilies of which the Family of the Anatidæ is composed, only seven are represented in North America, containing, according to the author's views, sixty-two species and subspecies, some of which, however, are not strictly natives of the continent, but merely stragglers within its borders. The first of the subfamilies, following the arrangement decided upon for this book, is:

SUBFAMILY CYGNINÆ.

THE SWAN.

IN this division are placed the largest birds among the Water Fowl, the Swan. There are but few species, and these are found pretty much throughout the world. Usually of an immaculate white plumage when adult, there is one exception, the Australian Swan, which is black, thus sustaining the character of the general fauna of that continent, in being different from those of other parts of the world. There are about eight species known of Swan or Swan-like birds, placed in three genera, five

confined to Europe, Asia, and North America, two to
South America, and one to Australia. The majority are
large birds with long, flexible necks, and some with
powerful voices, one only being mute. They associate
in flocks of from five or six to thirty, sometimes even
more, and are very conspicuous objects in the places
where they are accustomed to resort. Of this subfamily
only one genus is represented in North America.

GENUS CYGNUS

(Greek κύκνος, *kuknos;* Latin *cygnus*, a swan.)

Cygnus Bechst. Orn. Taschenb., 1803, vol. ii., p. 404 (note).
Type *Anas olor*, Gmel.

Bill as long as head, high at base, deeper than wide, broad
and rather flat at tip. Skin of bill reaching to eyes. Nostrils
situated high, and placed about the middle of the length of bill.
Neck very long and flexible. Tibiæ bare on lower part. Legs
behind center of body. Tarsus shorter than middle toe and
claw. Feet large. Wings long. Tail short.

For a long series of years the term CYGNUS, given by Bechstein, as recorded above, was adopted by all ornithologists
throughout the world for the White Swan. In 1832 Wagler
proposed the term *Olor*, which was a specific name for the
European Swan, but this was not generally, if at all, adopted by
naturalists. In 1882 Stejneger revived this term in his paper on
the CYGNINÆ, published in the Proceedings of the United States
National Museum, including in it two European species, *cygnus*
(*Anas cygnus*, Linn.) and *bewickii;* also two American species,
columbianus and *buccinator*. The only difference he mentions
in the diagnoses of the genera, *Cygnus* and *Olor* as given on
pages 189 and 197, is that the down on the head of the young in
Cygnus does not form distinct loral antiæ; but it does do this in
Olor, and also that the tail of the species of *Cygnus* is cuneate,
but rounded in *Olor*. These differences are all the characters
produced which are claimed as generic. In questioning the
wisdom or even the advisability of this attempt to reinstate *Olor*
as here formulated, and thus suppressing a term in which the

majority of Swan have been placed for many years (the reasons given being so very slight and insufficient), I am fully aware of the difficulties that exist in deciding as to what kind of characters and how many, in the conflicting opinions of ornithologists, there should be to properly establish a genus; for upon this subject there is not complete accord among naturalists. But, waiving these points, it is generally conceded that THE character or characters upon which a genus is founded should at least be permanent, so that an animal included in that genus might at all stages of its ADULT existence be able to exhibit the proofs that it properly belonged there. Otherwise, if this should not be so, a species, as it underwent modifications at different periods of its life, would have to be included in various genera, a proposition not to be entertained for a moment by any serious scientific person. The main character to separate *Cygnus* and *Olor* from each other, as given by Stejneger, is, as I have already quoted, the distribution of the down on the head of the *young* birds, an evanescent, adolescent, and unreliable distinction, one not possessed by the adults, and which, if recognized, would place the young in one genus, the adults in another. This fact is indisputable, and the error it embodies is one no ornithologist should countenance, much less perpetuate by any act of his own. The single remaining point, a cuneate or rounded tail, of itself can hardly be deemed sufficient to establish a genus, even by the most extreme advocate of novelties. For the reasons here given, which to my mind are ample, I have not adopted *Olor*, but have retained the familiar and appropriate term by which the White Swan have been so long known.

Three species of Swan are now included in the avi-fauna of North America; one, however, possessing but slight claims to be considered a resident of the continent. Of the two that are unquestionably North American, the Trumpeter has a comparatively restricted dispersion, and is not nearly so well known as its relative, the Whistling Swan. Both are magnificent birds, the Trumpeter, as its name implies, being remarkable for its sonorous voice. The Whooping Swan, a straggler into far-away Greenland, is a native of the Eastern Hemisphere, and has never appeared upon the continent of North America. It is easily recognizable, if anyone should happen to meet it within our boundaries (a very unlikely event), by the large amount of yellow on the bill.

KEY TO THE SPECIES.

A. Plumage of adults entirely white.
  a. Bill all black.

    a'. Yellow spot on lores near eye. — WHISTLING SWAN. *C. columbianus.*

    b'. No spot on lores. — TRUMPETER SWAN. *C. buccinator.*

  b. Bill, with basal portion and lores, yellow; remainder black. — WHOOPING SWAN. *C. cygnus.*

SUBFAMILY ANSERINÆ.

THE GEESE.

This subfamily includes the Geese of the world, arranged in six genera, possessing about twenty-five species. Geese are about halfway between the Swan and Ducks, having moderately long necks, rather long legs carrying the body well above the ground, and a comparatively easy, though not a graceful, walk. They are provided with a strong bill, and subsist largely upon grass, which they break off from the root by a quick jerk sideways. They have a powerful flight, capable of being sustained for many hours at a time, and the species are in the habit of associating in large flocks. The flesh is very palatable, especially that of the young birds, and in the Arctic regions these fowl are the main support of large numbers of people. While as a rule the different species are confined to separate continents or portions of continents, there are cases where the same species inhabits the northern part of both hemispheres. Many of them can be domesticated, and they will breed in confinement. Five genera of this subfamily are represented in North America.

KEY TO THE GENERA.

A. Lores feathered.
  a. Serrations on the greater portion of the cutting edges of maxilla visible.
    a′. Plumage all white, or head and neck only entirely white. Primaries, black or blackish brown.
      a″. Bill stout; depth at base more than half the length of the culmen; no excrescences on basal portion. Black space on commissure. Size large. } SNOW GEESE. *Chen*.
      b″. Bill weak; depth at base less than half the length of the culmen; basal portion covered with wart-like excrescences. No black space on the commissure. Size small. } ROSS'S SNOW GOOSE. *Exanthemops*.
    b′. Plumage never all white, nor with an entirely white head or neck. } LAUGHING GEESE. *Anser*.
  b. Serrations of cutting edge of maxilla visible only at angle of the mouth; commissure concave. } EMPEROR GOOSE. *Philacte*.
  c. Serrations of cutting edge of maxilla not visible; commissure straight. } CRAVAT GEESE. *Branta*.

GENUS CHEN

(Greek χήν, *chen*, a goose).

Chen, Boie. Isis, 1822, p. 563. Type *Anser hyperboreus*, Pall.

Bill as long as the head, powerful, higher than wide at base, edges of maxilla and mandible greatly beveled, exposing the prominent lamellæ. Nostrils situated high on basal portion of maxilla. Tarsus longer than middle toe and claw. Feet rather small.

Two species and one subspecies are retained in this genus, two of which, the Greater and Lesser Snow Geese, possess nothing to distinguish them apart save a difference in size. This, on an average, is stated to be about nine inches in the total length, but as there is a great variation in the measurement of individuals, it is not easy at times to determine as to which form an individual belongs. These two Snow Geese are distributed in their migrations over all North America, the imaginary dividing line of the species and subspecies being the Mississippi Valley, which is the winter locality of the less known Blue Wavey or Blue Goose.

KEY TO THE SPECIES.

A. Feathering on lateral base of maxilla, convex; blackish space at commissure.

   a. Plumage chiefly grayish brown and bluish gray. } BLUE GOOSE. *C. cærulescens*.

   b. Plumage all white save primaries, which are black.

   a'. Size small. Average total length said to be 28 inches. } LESSER SNOW GOOSE. *C. hyperboreus*.

   b'. Size large. Average total length said to be 34 inches. } GREATER SNOW GOOSE. *C. h. nivalis*.

GENUS EXANTHEMOPS

(Greek ἐξάνθημα, *exanthema*, eruption + ὄψις, *opsis*, resemblance).

Exanthemops, Elliot. B. of North America, 1868, vol. ii., pl. xliv., text. Type *Anser rossi*, Cass.

Base of bill thickly covered with wart-like excrescences; bill weak, no gape at commissure and no blackish space present. Feathering on lateral base of maxilla nearly straight. Size very small.

Only one species of this very distinct genus is known, the diminutive Ross's Goose. It is no larger than many species of Ducks, and can always be readily distinguished from all Geese, in addition to its small size, by the conspicuous and unusual excrescences at the base of the bill, which in some specimens cover this part entirely.

GENUS ANSER

(Latin *anser*, a goose).

Anser, Briss. Orn., 1760, vol. vi., p. 261. Type *Anas anser*, Linn.

Bill stout, not longer than head, depth at base less than half the length of culmen, tapering to tip. Serrations of maxilla visible when bill is closed. Nostrils on basal half of maxilla, placed high up near culmen. Tarsus shorter than middle toe and claw.

The White-fronted Geese of the Old and New Worlds have been separated as a species and subspecies on a difference of size averaging one inch in the total length of the adult and .37 inch in extent of the culmen. This is a worse case than the Snow Geese, because the White-fronted Geese of the two hemispheres are so nearly equal in their dimensions that, the locality of a specimen being unknown, its identification is impossible, for it would not be difficult to find individuals among the European White-fronted Geese that were even larger than some of the American.

As I have had occasion to remark, when writing of certain other species in this book, size alone is a most unsatisfactory character (?) to go by in determining species or subspecies, and when persisted in is most apt to create confusion.

In this instance I do not consider that this slight difference of dimensions is of sufficient consequence to cause the recognition of two forms of this Goose, and in this book, therefore, I have placed the species and its so-called subspecies under the name bestowed by Gmelin, and after careful study of the question, and examination of examples from both hemispheres, I should require better evidence than any yet produced to convince me that it is desirable to establish more than one form of this species.

GENUS PHILACTE

(Greek φῖλος, *philos*, loving + ἀκτη, *akte*, seashore).

Philacte, Bann. Proc. Acad. Scien., Phila., 1870, p. 131. Type *Anas canagica*, Sevast.

Bill stout, with the teeth exposed only at angle of the mouth. Nostrils situated on anterior end of the nasal fossæ. Nail prominent, occupying all the tip. Cutting edge of maxilla concave. Skull with superorbital depressions, an unusual character. Tarsus not longer than middle toe and claw. Webs of feet, excised.

But one species of this genus is known, an inhabitant of the Alaskan coasts, and some of the Aleutian and other islands in the Northwest, very occasionally straggling into the Pacific coast States of the Union. It is a very handsome Goose, rather heavy in body and of limited dispersion; a bird of the bleak regions of the north, never, unless by accident, penetrating into temperate climes. Great numbers are annually destroyed by the natives, and its probable extinction is not likely to be long delayed.

GENUS BRANTA

(Greek * βρένθος, *brenthos*, an unknown water bird).

Branta, Scop. Ann. I. Hist. Nat., 1769, p. 67. Type *Anas bernicla*, Linn.

Bill short, high at base; nostrils situated about the middle; serrations not visible; commissure straight. Feet rather small.

With the exception of one species, which is a straggler within our limits, all the members of this genus are natives of North America. It comprises the various forms of the "Cravat" or Common Wild Goose, and the smaller species known as Brant or Brent. They are scattered over the United States during the winter months, throughout its length and breadth, the various species having their own line of migration, which is rarely departed from, though a few, like the Canada Goose, are met with across the continent from ocean to ocean. Some of the species can be domesticated, bear confinement well, and will breed in captivity. The flesh of the young is very palatable, but that of the old birds is to be carefully avoided.

* If this derivation is correct, the proper name for the genus would be Brenthus and not Branta. But Brenthus was proposed by Schönherr in 1826 for a genus of Coleoptera, antedating Sundevall's employment of the same term (Meth. Nat. Av. disp. Tent., p. 145, 1873), and therefore it may not be used in ornithology. In case Branta therefore is not permissible, the next would be Leucoblephara, La Fres. 1840=Leucoblepharon Baird, 1858; each used by its author, however, as a subgenus. These failing. Leucopareia, Reichnb. Av. Syst. Nat., p. ix. (1852), is available. Bernicla (Boie Isis, 1822), is preoccupied (Bolt. crust. 1798).

KEY TO THE SPECIES.

A. Head black, cheeks white.
  a. General color dark brown, under parts light brownish gray, grading into white, no white collar at base of neck.
    a'. Size large. Average total length about 39 inches. Tail feathers, 18-20. — CANADA GOOSE. *B. canadensis.*
    b'. Size small. Average total length about 29 inches. Tail feathers, 14-16. — HUTCHINS' GOOSE. *B. c. hutchinsi.*
  b. General plumage light brown, under parts dark brownish gray, abruptly separated from white anal region. White collar sometimes at base of neck.
    a'. Size large. Average total length about 35 inches. Tail feathers, 18-20. — WHITE-CHEEKED GOOSE. *B. c. occidentalis.*
    b'. Size small. Average total length about 24 inches. Tail feathers, 14-16. — CACKLING GOOSE. *B. c. minima.*
  c. General plumage, bluish gray; under parts, grayish white. — BARNACLE GOOSE. *B. leucopsis.*
B. Head and cheeks all black.
  a. White patch on middle of neck, composed of streaks. — BRANT GOOSE. *B. bernicla.*
  b. Broad white collar on middle of neck, interrupted behind. — BLACK BRANT. *B. nigricans.*

The specimen of Hutchins' Goose mentioned in the article on that species as having been killed at Puckaway Lake, and now in the New York Museum of Natural History, is rather peculiar from the fact that while the under parts are light brownish gray, grading into the white of the anal region, and in this respect possessing the distinctive mark that separates its species and the typical Canada Goose from their allies, there is also a narrow white ring at the base of the neck; a character, at all events at certain seasons of the year, of *B. c. occidentalis* and *B. c. minima.* The style of the coloring on the under parts in the two divisions of these Geese is apparently much more to be relied upon than is the presence or absence of the white ring around the neck, and as this Puckaway example is unquestionably *B. c. hutchinsi*, from the coloring of the under parts, it is therefore evident that, occasionally, at all events, this subspecies assumes the white ring, as it does not seem at all necessary that the question of hybridism in this case should be considered, although it is true the bird was associating, at the time it was killed, not with its own fellows, but with a flock of Canada Geese. The white collar, however, would appear to be a rather doubtful character.

SUBFAMILY PLECTROPTERINÆ.

THIS subfamily comprises what I may call the Geese-like Ducks, with moderately short necks, rather long hind toe, not lobed, long tail, the feathers broad and rounded and with long upper coverts, and some like those in the following genus having short narrow bills high at base, tapering to the tip. It includes several genera but not all of them particularly related, some of the species having a rather brilliant plumage, with considerable metallic coloring. They are scattered all over the world; only one, however, being found in North America.

GENUS ÆX

(Greek αἴξ, αιχ, a water bird).

Aix; Boie (misspelling for Æx). Isis, 1828, p. 329. Type *Anas galericulata*, Linn.

Bill high at base, tapering toward tip, shorter than head or tarsus. Basal portion of maxilla forming a sharp angle between feathers of lores and forehead. Lamellæ small and few. Nostrils large, oval. Head crested. Tail feathers very broad and rounded at tip; rectrices sixteen, upper coverts very long. Tarsus shorter than middle toe.

Two species only are contained in this genus, the most beautiful of the Family, one of which, the Wood Duck, is a native of

North America and a doubtful straggler to the Old World; the other, the Mandarin Duck, confined to China, Formosa, and Japan. The Wood Duck of late years appears to be growing less plentiful, the beautiful plumage of the male causing it to be a desirable object for various purposes, one of which is dressing artificial flies, the exquisite flank feathers being especially selected for that purpose.

SUBFAMILY ANATINÆ.

FRESH-WATER DUCKS.

This is one of the great divisions of the family and contains what may with a certain degree of propriety be called the Fresh-Water Ducks, though it must not be understood from that term that none of the species ever go to the sea. They are readily distinguished from the members of the subfamily FULIGULINÆ or Salt-Water Ducks by the shape of the hind toe, those of the ANATINÆ having that member simple or normal, the FULIGULINÆ having it lobed or flat. The River or Fresh-Water Ducks have moderately short necks and legs, excepting Dafila and Dendrocygna, while the feet are much smaller than those of the Sea Ducks. As a rule they are poor divers, and procure their food mostly in shallow water, by tilting the hinder part of the body so that they can reach the grasses, etc., growing on the bottom and pull it up with their bills. When wounded they skulk, laying the head and neck flat upon the water, and seek the nearest marsh for concealment. The flesh of these Ducks is generally most palatable, the exceptions being those individuals that may associate and feed even temporarily with the Sea Ducks, when they have usually a very fishy flavor. These birds moult twice a year, and the sexes are dissimilar in plumage.

KEY TO THE GENERA.

(For North American Species.)

I. Hind toe not lobed.

A. Lower part of tarsus in front, without transverse scutellæ. Neck and legs long. — TREE DUCKS. *Dendrocygna.*

B. Lower part of tarsus in front with transverse scutellæ.

   a. Bill not spatulate.

      a'. Lamellæ of mandible projecting outward. — RUDDY SHELDRAKE. *Casarca.*

      b'. Lamellæ of mandible not projecting outward.

         a''. Bill, depth at base less than width; broader toward tip than at base. — MALLARDS. *Anas.*

         b''. Bill, depth at base equal to width; narrower at tip than at base.

            a'''. Lamellæ of maxilla prominent. Central tail feathers not elongated. — GADWALL. *Chaulelasmus.*

            b'''. Lamellæ of maxilla moderate. Central tail feathers moderately elongated — WIDGEON. *Mareca.*

         c''. Bill, depth at base greater than width. Sides of maxilla nearly parallel.

            a'''. Central tail feathers much elongated. Neck very long. — SPRIGTAIL. *Dafila.*

            b''''. Central tail feathers not elongated. Neck short.

                a⁴. Upper wing coverts blue or bluish gray. — BLUE-WINGED TEALS. *Querquedula.*

                b⁴. Upper wing coverts brownish or slaty gray. — GREEN-WINGED TEALS. *Nettion.*

   b. Bill spatulate. — SHOVELER. *Spatula.*

GENUS DENDROCYGNA

(Greek δένδρον, *dendron*, a tree + Latin *cygnus*, a swan).

Dendrocygna, Swain. Class. B., 1837, vol. ii., p. 365. Type *Anas arcuata*, Cuv.

Bill as long as the head, nail occupying nearly all the tip, and curving downward. Nostrils ovate, situated high upon the bill and on the basal portion. Neck long and slender. Legs very long, lower part of tibiæ denuded; lower portion of tarsus in front without transverse scutellæ, but covered with small scales, like those of Geese. Hind toe one-third the length of tarsus.

There are about nine Tree Ducks belonging to this genus, scattered over various portions of the world. They are peculiar for their long legs and necks and have affinities for Geese. They roost and nest in trees, and have a variously colored plumage, some species being very attractive. Two only are found in North America, penetrating into the States along our southern border, for these Ducks are chiefly dwellers in tropical lands.

KEY TO THE SPECIES.

A. Brownish black stripe down hind neck.

   a. Abdomen and flanks black. } BLACK-BELLIED TREE DUCK. *D. autumnalis.*

   b. Abdomen and flanks cinnamon. } FULVOUS TREE DUCK. *D. fulva.*

GENUS CASARCA

(Russian *cacharca*, sea swallow).

Casarca, Bon. Comp. List B. Eur. and Amer., 1838, p. 56. Type *Anas casarca*, Linn.

Bill with parallel sides, culmen nearly straight, lamellæ of mandible projecting outwardly. Lower portion of tarsus in front with transverse scutellæ. Tarsus rather long.

A genus containing four handsome species, very goose-like in their habits and in the tones of their voices. They are essentially birds of the Old World, and although two examples of one species are supposed to have accidentally straggled into Greenland, that fact is hardly sufficient to give it a rightful claim to be included among North American birds, especially as I am not aware that those who record its presence in Greenland saw the individuals there alive.

GENUS ANAS

(Latin *Anas*, a duck).

Anas, Linn. Syst. Nat., 1766, vol. i., p. 134. Type *Anas boschas*, Linn.

Bill about as long as the head, longer than the tarsus, broad and swelling outward toward the tip, where its greatest width is nearly one-third the length of the culmen.

In the A. O. U. Check List this genus is made to include a number of species such as the Gadwall, Widgeon, and Teal, in addition to those closely related to the type. Genera, of course, are not found in nature, but afford convenient boundaries for the more complete arrangement of groups in natural science. So perhaps it would not be absolutely incorrect if all the Fresh-Water Ducks were placed under ANAS; but as a number of them possess characters which may properly be called generic, and which are not possessed by others, there is no reason why these should not be recognized. To be consistent we must do one of two things: include most of the species under one genus, or accept the fact that there are numerous genera and recognize the characters that indicate them wherever found. ANAS, as I regard it, possesses only four species and subspecies in North America, one (*A. f. maculosa*) possibly of doubtful validity, as we become more familiar with its claims for separation from the others. There are nearly twenty species that belong to this genus, not including any of those not typical retained in it according to the A. O. U. List, but which properly should be placed in other genera. These twenty species are scattered throughout the world, and from the type, the Common Wild Duck, are descended most of the domesticated races. The members of this genus rarely go beyond the Arctic circle, and the species often remain in the temperate zone throughout the year, and breed wherever they may be. In fact, the two subspecies inhabiting the United States are rarely met with as far north as Kansas. They are "mud ducks"; that is, fond of dabbling in the ooze found along the banks of streams or the bottoms of shallow creeks, and obtain most of their food by sifting the liquid mud through the lamellæ of the bill.

KEY TO THE SPECIES.

A. Central upper tail coverts of male re-curved. White on wing coverts. — MALLARD. *A. boschas.*

B. Central upper tail coverts of male not re-curved. No white on wing coverts.

   a. Sides of head and throat, grayish fulvous, closely streaked with black. — DUSKY DUCK. *A. obscura.*

   b. Sides of head and throat, pale buff, sometimes streaked with black on cheeks and portions of neck. — FLORIDA DUSKY DUCK. *A. fulvigula.*

   c. Sides of head and throat, buff, streaked with black. Under parts, mottled with buff and blackish brown. — MOTTLED DUCK. *A. f. maculosa.*

GENUS CHAULELASMUS

(Greek χαύλιος, *chaulios*, protuberant + ἔλασμος, *elasmos*, a plate).

Chaulelasmus, G. R. Gray. Bon. Consp. List, B. Eur. and N. Am., 1838, p. 56. Type *Anas strepera*, Linn.

Bill about two-thirds length of head, longer than tarsus, slender, widest at base, greatest width less than one-half the length of culmen. Lamellæ of maxilla prominent. Tail pointed, median rectrices not elongated.

This genus contains the well-known Gadwell or Creek Duck, a cosmopolitan species of the Northern Hemisphere, and possibly a smaller form inhabiting the Fanning Islands in the Pacific Ocean. The male is readily distinguished from other species of North American Ducks by having a great deal of chestnut color on the wing coverts, and the female by her gray and white speculum.

GENUS MARECA

(Mareca. Brazilian name for Teal).

Mareca, Steph. Gen. Zoöl., vol. xii., 1824, pt. ii., p. 130. Type *Anas penelope*, Linn.

Bill small, tapering toward the tip, nearly half as long as head, and about equal in width throughout its length. Central rectrices moderately lengthened.

Two species, out of the three known to belong to this genus, are found within our borders; one indigenous to the Continent, the other a frequent straggler from the Old World. Both are beautiful birds, the male's plumage being gayly colored, but the two forms have little or no resemblance to each other. The European Widgeon has more strongly contrasted colors perhaps than those seen in its American relative, but neither has very much advantage over the other in beauty. The exotic species has been taken many times in various portions of the United States, all males, however; the female, having such a close resemblance to that of our Baldpate, would probably pass unnoticed, even if captured.

KEY TO THE SPECIES.

A. Top of head buff, rest of head and neck chestnut. } EUROPEAN WIDGEON. *M. penelope.* ♂

B. Top of head whitish; rest of head and neck whitish, spotted with black, and with a lengthened patch of metallic green. } BALDPATE. *M. americana.* ♂

C. Head and upper neck reddish brown, spotted with black. } EUROPEAN WIDGEON. *M. penelope.* ♀

D. Head and upper neck whitish, spotted with black. } BALDPATE. *M. americana.* ♀

GENUS DAFILA

(Dafila, nonsense word).

Dafila, Steph. Gen. Zoöl., vol. xii., pt. ii., 1824. p. 126. Type *Anas acuta*, Linn.

Bill long as head, slender, the width about one-third the length of culmen, and nearly equal throughout; neck very long and slender. Central rectrices greatly elongated. Wing pointed. First and second primaries equal and longest. Feathers of lores form a convex line at base of maxilla.

This genus contains only three species, widely separated: one the North American, which, however, is found also throughout the northern part of the Northern Hemisphere; one from South America; and one from Kerguelen Island. The American Sprigtail is a gracefully formed bird, and although its neck may seem disproportionately long, it does not appear so when the bird is quietly swimming along intently seeking its food. The Sprigtail is mainly a fresh-water Duck, and although it is found on the sea-coast, yet even there it seeks the bays and sounds where the water is brackish. It goes at times in large flocks, and consorts frequently with the Widgeon, the two species flying about together.

GENUS QUERQUEDULA

(Latin *Querquedula*, a kind of Teal).

Querquedula, Steph. Gen. Zoöl., vol. xii., pt. ii., 1824, p. 142. Type *Anas querquedula*, Linn.

Bill about as long as head, longer than tarsus; narrow, sides parallel; greatest width more than one-third length of culmen. Tail pointed. Head not crested.

Two of the four species belonging to this genus are found in North America. The males in full dress are very handsome birds and strikingly different in the color of their plumage from other Ducks. The habits of these teal and those of the genus Nettion are very similar. Both go in flocks of considerable size, have a swift, erratic flight, resort to like localities and seek the same kind of food. In addition to its attractive appearance, the Blue-winged Teal is one of our very best table birds, the flesh being tender and juicy, and when it has been feeding upon wild rice, is then of exceptionally fine flavor.

KEY TO THE SPECIES.

A. Head and neck dull plumbeous. White crescentic patch between eye and bill. } BLUE-WINGED TEAL. *Q. discors.* ☦

B. Head and neck bright chestnut. No white patch between eye and bill. } CINNAMON TEAL. *Q. cyanoptera.* ☦

C. Throat and abdomen white.	BLUE-WINGED TEAL. *Q. discors*. ♀
D. Throat deep buff. Abdomen rufous, mottled with black.	CINNAMON TEAL. *Q. cyanoptera.* ♀

GENUS NETTION

(Greek νέττιον, *nettion*, a duckling, dim. of νέττα, *netta*, a duck).

Nettion, Kaup. Natürl. Syst., 1829, p. 95. Type *Anas crecca*, Linn.

Bill two-thirds as long as the head, much longer than tarsus, slender, slightly narrowing toward the tip; greatest width one-third the length of culmen. Head not crested.

About a dozen species of this genus are distributed throughout the world, of which only one is indigenous to North America. The European Green-winged Teal, a close ally and easily confounded with the North American species, occasionally straggles into our limits, perhaps more frequently than is supposed, as the ordinary observer would not notice any difference between them. Both species go in flocks of considerable size, and have a swift, erratic flight.

KEY TO THE SPECIES.

A. A broad crescentic white band in front of wing on either side of breast. No white on scapulars.	AMERICAN GREEN-WINGED TEAL. *N. carolinensis.* ♂
B. No crescentic white band in front of wing. Scapulars margined with white or buffy white.	EUROPEAN GREEN-WINGED TEAL. *N. crecca.* ♂

There appear to be no characters for distinguishing the females of the two species from each other.

GENUS SPATULA

(Latin *spatula*, dim. of *spatha*, a broad blade).

Spatula, Boie. Isis, 1822, p. 564. Type *Anas clypeata*, Linn.

Bill longer than head, spreading out toward the tip, where it is twice as wide as at the base. Nail prominent, forming a hook. Lamellæ prominent. Wings long, pointed. Tail short, composed of fourteen acute feathers. The peculiarly shaped bill makes this species readily recognizable among our Ducks, irrespective of other characters.

The Shoveler is cosmopolitan, and the American bird is one of the four known species of the genus. The others are natives of South America, Australia and its neighboring islands, and South Africa, respectively. Of the North American species, when arrayed in all its finery, the male is a beautiful bird, although, from the disproportionate size of the bill, it is apparently slightly top-heavy. It has, however, a graceful shape, and walks easily and well. The female, of course, can be distinguished from those of other species by her large, spoon-shaped bill.

SUBFAMILY FULIGULINÆ.

SEA DUCKS.

This subfamily contains the Sea Ducks, which are mainly distinguished from the species of ANATINÆ, frequenting the Rivers and Lakes, usually known as the Fresh-Water Ducks, by having a membranous web depending from the hind toe. The feet are larger, with broader webs and longer toes, while the legs are shorter and placed nearer the tail, causing the walk to be awkward and somewhat difficult, but facilitating both swimming and diving. Most of the species belong to the Northern Hemisphere, and breed in high latitudes, and a large number are exclusively marine, but others are seen occasionally on the Great Lakes and large rivers. Individuals found in such localities are, however, usually young birds, which probably from either fatigue or hunger have made a brief stop while migrating.

The members of this subfamily are great divers and subsist upon mollusks, fish, various grasses, and bulbous roots which they procure on or near the bottom. Their flesh varies greatly according to the kind and quality of their food; those subsisting upon a fish diet possess often an " ancient fish-like smell " and taste, while those that feed on leaves, or roots of the more delicate plants, such as the wild celery, are very tender and of excellent flavor. The sexes are usually very different in the hues of their plumage, the principal exceptions to this being among the Scoters of the genus ŒDEMIA. There is much diversity of structure among these birds, necessitating quite a number of genera, and the specific characters are strongly marked, and consequently easy of recognition. These Ducks feed mostly by night, the persecutions to which they are subjected preventing them from obtaining their food during the day, at which time, weather permitting, they assemble in large numbers in the middle of broad waters and sleep or dress their feathers. Moonlight nights are favorite ones for feeding, and on such occasions they visit creeks or ponds in marshes near the sea. The notes uttered by these birds are harsh and guttural, and the animated, inspiring *quack* of some of the fresh-water species is never heard among them.

KEY TO THE GENERA

(*For North American Species.*)

A. Hind toe broadly lobed.

    a. Head with an elongated crest.   } RUFOUS-CRESTED DUCK. *Netta.*

b. Head without elongated crest.

    a'. Bill long as middle toe without claw, greatest width LESS than one-third the length of culmen. Head long, not bunchy. — CANVAS BACK. *Aristonetta.*

    b'. Bill shorter than middle toe without claw, greatest width MORE than one-third the length of culmen. Head bunchy. — RED-HEAD. *Æthyia.*

    c'. Bill shorter than head, broad, greatest width nearly HALF the length of the culmen. — SCAUP DUCKS. *Fuligula.*

    d'. Bill with membranous expansion on edge of maxilla near tip. — LABRADOR DUCK. *Camptolæmus.*

    e'. Bill very short, narrow, rather pointed.

        a''. Anterior edge of nostrils nearer the base than tip. — GOLDEN EYE DUCKS. *Clangula.*

        b''. Anterior edge of nostrils nearer the tip than the base. — BUFFEL HEAD DUCK. *Charitonetta.*

        c''. Bill, height at base two-thirds length of culmen.

            a'''. Central tail feathers elongated. — LONG-TAILED DUCK. *Havelda.*

            b'''. Central tail feathers not elongated.

                a^4. Bill shorter than tarsus. — HARLEQUIN DUCK. *Histrionicus.*

                b^4. Bill longer than tarsus. — STELLER'S DUCK. *Heniconetta.*

    f'. Bill tumid or gibbous. — SURF DUCKS. *Œdemia.*

    g'. Bill with two-thirds of the culmen covered with feathers; pad-like feathering around the eyes. — FISCHER'S EIDER DUCK. *Arctonetta.*

    h'. Bill with naked parallel frontal processes. Feathering around eyes normal. — EIDER DUCKS. *Somateria.*

GENUS NETTA

(Greek νέττα, *netta*, a duck).

Netta, Kaup. Naturl. Syst., 1829, p. 102. Type *Anas rufina*, Pallas.

Bill broadest at the base, narrowing gradually toward the tip; nail broad and prominent, more than one-third the width of the bill. Outline of loral feathering slightly concave. Culmen longer than tarsus. Head of male with lengthened crest.

One species only, the Rufous-crested Duck, is contained in this genus, an inhabitant of the Old World, where it ranges from the basin of the Mediterranean to Turkestan and Northern India, only casual in Northern Europe and Great Britain. In North America I am not aware that anyone has ever seen it alive, and even as a straggler it has little claim to a place in our avi-fauna.

GENUS ARISTONETTA

(Greek ἄριστος, *aristos*, best + νέττα, *netta*, a duck).

Aristonetta, Baird. B. N. Am., 1858, p. 793. Type *Anas valisneria*, Wils.

Bill as long as middle toe without claw; longer than head; greatest width less than one-third the length of the culmen, greatly depressed toward tip; nail moderate, not hooked. Culmen depressed in center for nearly one-third the length of bill from base. Head long, not bunchy; neck of equal diameter throughout its length.

A comparison of the above diagnosis with that of the one succeeding gives ample evidence of the generic distinction of the Canvas Back and Red-Head, and I do not consider that such radical differences as are to be observed between the two species can be properly accentuated by the employment of ARISTONETTA subgenerically. The Red-Head has numerous and some very close allies throughout the world having the same generic characters, while the Canvas Back is *sui generis*, and has no exotic representatives nor home relatives. Its very peculiar bill and thick neck, the latter of nearly equal diameter for its entire length, cause it to be conspicuous among the Duck tribe and without imitators, unless the small Ruddy Duck, with its thick neck, can be considered as such.

GENUS ÆTHYIA

(Greek αἴθυια, æthyia, a sea bird).

Aythya (misspelling for Æthyia), Boie. Isis, 1822, p. 564. Type *Anas ferina*, Linn.

Bill shorter than middle toe without claw, as long as head, the greatest width more than one-third the length of the culmen. Height of maxilla at base equal to its greatest width, moderately depressed toward tip. Nail prominent and hooked. Head bunchy, larger than neck, which is compressed at the throat.

There is only one species in North America belonging to this genus, the well-known Red Head, as the Canvas Back, which has usually been placed in it, I regard as generically distinct. The genus, however, is represented in South America, and also in the Old World from Great Britain to Japan as well as in Africa, Australia, and some of the contiguous islands. One Old-World species, *Æ. ferina*, resembles very closely the American bird, and when on the water might be mistaken for it.

GENUS FULIGULA

(Latin *Fulica* or *Fulix*, a coot, dim. *fulicula*, or possibly, dim. of *fuligo*, soot, black.

Fuligula, Steph. Gen. Zoöl., vol. xii., pt. ii.; 1824, p. 187. Type *Anas fuligula*, Linn.

Bill short, broad, not as long as head, widest at tip, greatest width nearly half the length of culmen, moderately depressed, with a broad nail terminating in a hook. Height of maxilla at base less than greatest width. Tarsus little less than half the length of middle toe and claw. Head bunchy, neck rather slender.

Three species of this genus are found in North America, one of which, the Big Black Head, *F. marila*, is also a native of the Eastern Hemisphere. The specimens of this species obtained within our boundaries have been separated from those of the Old-World by American ornithologists, but the characters relied upon to distinguish the two forms are not apparently tenable, the American examples, even among those shot in one locality, as was clearly shown by Mr. Bishop (Auk, 1895, p. 293), exhibiting the differences, with gradations, that were attributed to the two birds. The question, therefore, as to whether there is both a distinct species and sub-species of the Big Black Head would seem to be clearly settled in the negative.

KEY TO THE SPECIES.

A. No ring around neck.
  a. Head and neck black, glossed with metallic green. } BIG BLACK HEAD. *F. marila.* ♂
  b. Head and neck black, glossed with metallic purple. } LITTLE BLACK HEAD. *F. affinis.* ♂

B. Ring around neck. } RINGED-NECK DUCK. *F. collaris.* ♂

C. White patch on wing.
  a. Length of wing 8¼ inches or over. } BIG BLACK HEAD. *F. marila.* ♀
  b. Length of wing 8¼ inches or less. } LITTLE BLACK HEAD. *F. affinis.* ♀

D. Bluish gray patch on wing. } RINGED-NECK DUCK. *F. collaris.* ♀

GENUS CAMPTOLÆMUS

(Greek καμπτός, *kamptos*, flexible; + λαιμός, *laimos*, throat).

Camptolæmus, G. R. Gray. List. Gen. B. ed. 2, 1841, p. 95. Type *Anas labradorius*, Gmel.

Bill about as long as head, very broad, height at base not equal to greatest width. A membranous expansion, on the edge of maxilla toward the tip, increases considerably the normal width of the bill. Nail prominent, forming a hook at tip. Nostrils oblong, basal, and situated rather high on the side of maxilla. Loral and cheek feathers stiff, with horny tips, extending on to base of maxilla in a convex line. Tail of fourteen feathers, short.

The single, rather peculiar species, comprising this genus, while very common on certain parts of our eastern seaboard fifty years ago, is now extinct. It was remarkable for the unusual structure of the bill, which differs from all those of living species of Ducks, and for its striking black and white plumage. It was a strong flyer, and apparently perfectly competent to take care of itself, and the cause of its disappearance from our Continent is an unfathomable mystery. Many theories have been advanced to account for its extinction, but, as none admit of proof, it is impossible to arrive at a satisfactory explanation.

GENUS CLANGULA

(Latin *clangula*, dim. of *clangor*, a noise).

Clangula. Leach in Ross, Voy. Disc., App., 1819, p. xlviii. Type *Anas clangula*, Linn.

Bill shorter than head, high at base and tapering to tip. Nail prominent and hooked. Anterior end of nostril nearer to the tip than to the loral feathers. Tail rounded, of sixteen feathers.

Two species of this genus are found in North America, both of which are also natives of parts of the Eastern Hemisphere. The Common Golden Eye of our coasts and rivers, while in plumage it resembles in every particular the bird obtained in the Old World, has been separated as a distinct race, on account of being slightly larger on the average. A species or a race founded solely upon the slight, constantly varying size of individuals has a very difficult position to maintain in any family of birds, but is of a still more uncertain quantity when the establishment of so important a distinction is attempted in a like manner with members of the ANATIDÆ, as they notoriously vary in size, so that individuals of the same species can be readily found whose measurements differ at times in a surprising degree. It is only necessary to look at the measurements of a series of almost any species of the Anatidæ to see how wide apart the two extremes are, and within the range some examples would undoubtedly be found agreeing exactly with their foreign relatives, if they had any. It seems as if ornithologists acted at times under the conviction that, because a species is found in North America, it must be specifically or racially different from its Old-World representatives, and then the slightest variation is deemed sufficient to bestow upon it a new name. There are a number of such instances among the ANATIDÆ, which serve not only no useful purpose whatever, but mystify and confuse the student. Therefore, as I can find no reliable characters to distinguish the American and European Golden Eye from each other, and no certain line of demarcation between them, I have deemed it both unnecessary and unwise to retain the name given to our bird, for I cannot see that its claim to be considered even a subspecies has in any way been satisfactorily established.

KEY TO THE SPECIES.

A. Bill high at base, narrowing toward tip.
  a. Nostrils nearer the tip than base of bill.
    a'. Head and upper neck metallic green.   GOLDEN EYE. *C. clangula.* ☿
    b'. Head and upper neck metallic blue.   BARROW'S GOLDEN EYE. *C. islandica.* ☿
    c'. Head and upper neck hair brown.
      a''. Height of bill at base LESS than distance from anterior edge of nostril to nearest loral feathers.   GOLDEN EYE. *C. clangula.* ♀
      b''. Height of bill at base EQUAL to distance from anterior edge of nostril to nearest loral feathers.   BARROW'S GOLDEN-EYE. *C. islandica.* ♀

GENUS CHARITONETTA

(Greek χάρις, *charis*, graceful + νέττα, *netta*, a duck).

Charitonetta, Stejn. Bull. U. S. Nat. Mus., 1885, No. 29, p. 163. Type *Anas albeola*, Linn.

Bill about two-thirds length of head, height at base half the length of culmen. Nail rather narrow, curving downward. Anterior end of nostril nearer the loral feathers than the tip of bill. Head bunchy. Tail more than twice as long as tarsus.

Only one species is included in this genus, the common Buffle Head Duck. By some authors it is kept in the previous genus Clangula. The Buffle Head is a native of North America, straggling occasionally, when it loses its way, to Cuba and even to Europe; the last, however, rather exceptional. The male is a beautiful bird, the head rejoicing in rich metallic colors, and in its general appearance he is a diminutive Golden Eye.

GENUS HAVELDA

(Havelda, Norw. *Havelde*, a Sea Duck).

Harelda (misprint or misspelling for Havelde). Stephens in Shaw's Gen. Zoöl., 1824, vol. xii., pt. ii., p. 174. Type *Anas glacialis*, Linn.

Bill shorter than head, equal to tarsus, widest at base, narrowing rapidly to tip. Nail hooked. No lateral angles from base of culmen, loral feathering at base of bill nearly a straight line. Nostrils situated high on basal half of bill. Tail pointed, of 14 feathers; median pair slender and greatly elongated.

Only one species is recognized of this genus, a native of both the Western and Eastern Hemispheres, the familiar Old Squaw, or South Southerly of sportsmen. The male is remarkable for the greatly elongated middle feathers of the tail. It is a Sea Duck, flesh fishy and disagreeable in flavor, goes in flocks of considerable size, and flies with great rapidity. There is a striking difference in the plumage of summer and winter, the male, especially, in the two seasons appearing like quite another bird.

GENUS HISTRIONICUS

(Latin *histrionicus*, theatrical, relating to the bird's fantastic coloring).

Histrionicus, Less. Man. d'Orn., 1828, vol. ii., p. 415. Type *Anas histrionicus*, Linn.

Bill small, about half the length of head, shorter than tarsus, tapering rapidly to the tip, which is rounded and occupied by the hooked nail. Height at base equal to the extreme width. Loral feathering convex on base of bill. Frontal feathers advancing on culmen beyond the lores. Nostrils basal, and situated high on bill just beneath the culmen. Tail pointed.

This handsome bird with its fantastic markings, known as the Harlequin Duck is the only species of this genus. The female is attired very differently from the male, and, by the side of her brilliant "Lord," she is a very plain little body. This species is essentially a bird of the north, rarely entering the waters of temperate climes, and while it has a wide distribution over northern North America, it is also a native of Iceland, straggling occasionally into European boundaries. The Harlequin, in some of its characters, leans toward the Eiders, with which the intervening genera help to connect it.

GENUS HENICONETTA.

(Greek ἑνικός, *henikos*, singular + νέττα, *netta*, a duck).

Eniconetta (aspirate ignored), G. R. Gray. List. Gen. B., 1840, p. 75. Type *Anas stelleri*, Pall.

Bill without frontal processes; height at base slightly more than greatest width, this last not quite equal to half the length of culmen, which is longer than tarsus. Sides of maxilla tapering gradually toward the tip, which is nearly all occupied by the nail. Nostrils ovate, basal, placed high on maxilla. Outline of loral feathering convex. Speculum on wing.

One species only of this genus is known, the beautiful Steller's Duck, a dweller in high northern latitudes. It gathers at times in great flocks in the desolate regions it frequents, and often associates with other Eiders inhabiting the same localities. This genus is sometimes spelled *Eniconetta*; but as this entirely ignores the aspirate of the Greek ἑ, and is therefore quite incorrect, I have not continued the error.

GENUS ŒDEMIA.

(Greek οἴδημα, *oidema*; Latin *œdema*, a swelling).

Oidemia (misspelling for Œdemia), Fleming. Phil. of Zoöl., vol. ii., 1822, p. 260. Type *Anas nigra*, Linn.

Bill variously tumid or gibbous; frontal feathers extending further on the bill than those of the loral region. Maxilla extending anterior to nostrils, thence narrowing rapidly to tip. Nail broad, occupying the entire tip, curved and hooked. Nostrils situated about middle of bill. Extreme width of bill greater than height of maxilla at base.

The Surf Ducks or Scoters, as they are frequently called, are very numerous on our coasts in winter. Four species inhabit North America, and while their plumage is somber, the males being either all black, or black and white, the bills of this sex in the different species are decorated with red, orange, or other brilliant colors. It is a cosmopolitan genus, the members being found in both hemispheres in northern latitudes. A fifth species, *Œ. carbo*, Pall., may possibly occur in Alaska, its proper habitat being Northeastern Asia, but as yet no specimens have been procured within the boundaries of North America. One species, the Velvet Scoter, attributed to the New World, is really a native of the Eastern Hemisphere, and only claims a place in our avifauna by the accidental appearance of individuals in Greenland, evidently stragglers from the regular route during migration. The flesh of these Ducks is tough and fishy, to be carefully avoided whenever served at table.

KEY TO THE SPECIES.

A. Maxilla more or less swollen at base.
   a. Plumage of male black with white patches on front and back of head. Black spot on swollen base of maxilla. — SURF SCOTER. *Œ. perspicillata.* ♂
   b. Entire plumage of male deep black.
      a′. No speculum. — AMERICAN SCOTER. *Œ. americana.* ♂
      b′. Speculum white.
         a″. Swollen lateral basal part of maxilla bare. — VELVET SCOTER. *Œ. fusca.* ♂
         b″. Swollen lateral basal part of maxilla feathered. — WHITE-WINGED SCOTER. *Œ. deglandi.* ♂
         c″. A white spot at base of maxilla and one near ear.
            a‴. Upper parts brownish gray. — VELVET SCOTER. *Œ. fusca.* ♀
            b‴. Upper parts sooty brown. — WHITE-WINGED SCOTER. *Œ. deglandi.* ♀

GENUS ARCTONETTA

(Greek ἄρκτος, *arktos*, a bear + νῆττα, *netta*, a duck).

Arctonetta, G. R. Gray. Proc. Zoöl. Soc., 1855, p. 212. Type *Fuligula fischeri*, Brandt.

Bill rather small and narrow, with only a little over one-third of the culmen exposed, the rest covered by a mass of dense velvety feathers that come to a point beyond the nostrils which are partly hidden beneath them. From the culmen these feathers pass obliquely downward to edge of maxilla, and then backward to the end of the mouth. A line of feathers extends from chin on mandible nearly as far forward as those on the culmen. Nail occupying most of the tip, but there is no hook. Tertials falcate. Tail rounded, feathers inclined to a point.

One species represents this genus—the curiously marked Fischer's or Spectacled Eider of the northwest coast of America. It is common enough in the localities it frequents, but rarely comes to the southward of Alaska, and is pre-eminently a bird of the Arctic regions.

GENUS SOMATERIA

(Greek σῶμα, *soma*, body + ἔριον, *erion*, wool).

Somateria, Leach in Ross' Voy. Disc., app., 1819, p. xlviii. Type *Anas mollissima*, Linn.

Culmen about half as long as head. Bill slender with acute or rounded lateral, nearly parallel, processes reaching on the forehead between the extension of the frontal feathers and those on the sides, the former of which go nearly to the nostrils. Sides of bill tapering to the tip. This is entirely covered by the nail, which extends downward over the mandible when the bill is closed. Nostrils situated just in advance of the lateral feathering on the maxilla. Tertials curved downward over the wing.

This genus contains four well-characterized species, distributed in the Arctic regions of the Northern Hemisphere. Some are celebrated for their down, which is collected during the breeding season from the nests, and is an important article of commerce. It is plucked by the female from her breast to serve as a protection to the eggs. Three of the species are closely related, but the fourth, the King Eider or King Duck, differs in having a large squarish frontal process near the base of the bill. This, if permanent, would perhaps necessitate the removal of the species to a separate genus, but as it only exists during the breeding season, and at all other times the bill does not materially differ in outline from those of the other Eiders, the species is properly retained in the same genus with them. A subgeneric term, *Erionetta* (ἔριον, *erion*, wool + νῆττα, *netta*, a duck), was proposed for the King Eider by Coues in 1884.

The Eider Duck of the Old World, and the one obtained in Greenland, have been separated by American ornithologists for the same insufficient reasons given in similar cases of certain Geese and Ducks, viz., a slight difference in size, to which in this instance is added a variation in the color of the bill, "olive yellowish" instead of "olive green"; * a distinction, to most persons, practically without a difference. These characters, upon which a specific or subspecific separation of the birds is based, are not apparent to the ordinary observer, and only occasionally to the expert, and can hardly be deemed of sufficient importance, considering how Ducks vary in size, and also the difficulty of recognizing delicate distinctions of slight shades of olive, to require the Greenland and European birds to assume any kind of separate rank. Species or subspecies, where the individuals require a pair of dividers, or a great ability on the part of the investigator, to recognize intimately related shades of color for their maintenance, should not be permitted to obtain recognition in what ought to be regarded as a serious scientific study, for the differences are too apt to mislead, and seriously confuse and discourage the conscientious student.

KEY TO THE SPECIES.

A. Feathers of forehead reaching about half as far on bill as the loral feathers.

   a. Frontal angles on bill broad with round ends. } AMERICAN EIDER. *S. dresseri.*

   b. Frontal angles on bill narrow with pointed ends.

      a'. Without V-shaped mark on throat of male. } COMMON EIDER.† *S. mollissima.*

      b'. With black V-shaped marks on throat of male. } PACIFIC EIDER. *S. v.-nigrum.*

B. Feathers of forehead reaching to posterior end of nostril. } KING EIDER. *S. spectabilis.*

---

* Ridgway, Manual, 2d ed., 1896, p. 109.

† In some male specimens a dusky V-shaped mark is seen on the throat, but this is very exceptional.

SUBFAMILY ERISMATURINÆ.

SPINE-TAIL DUCKS.

THIS subfamily is represented throughout the world by many species comprised in about four genera, two of which are represented by only one species each in North America. One of these has a wide distribution within our limits, but the other can only be regarded as a straggler from more southern latitudes. They are peculiar little Ducks, with large heads, and very broad bills and feet, and the tail is composed of 18 to 20 stiff, pointed feathers, frequently carried directly upward. The males of both species have a brilliantly colored plumage, of red and black hues mainly, but this is only assumed by the resident bird during the breeding season. Both kinds are skilful divers, and fly with great rapidity, buzzing through the air more in the manner of insects than of birds. The flesh of these Ducks is fairly good, and of late years, probably from the growing scarcity of more desirable varieties, the Ruddy Duck has taken a rather prominent position in the markets of our land.

KEY TO THE GENERA.

(For North American Species.)

A. Tail feathers stiff, narrow, pointed.
   a. Nail of bill with the point bent downward and backward. } RUDDY DUCK. *Erismatura*.
   b. Nail of bill with the point perpendicular. } MASKED DUCK. *Nomonyx*.

GENUS ERISMATURA

(Greek ἔρισμα, *crisma*, a prop + οὐρά, *oura*, tail).

Erismatura. Bon. Sagg Distr. Met. Agg. e Corr., 1832, p. 143. Type *Anas jamaicensis*, Gmel.

Bill about as long as head, broad, widening toward the tip and turned slightly upward; nail small, narrow, curved, and turned backward. Nostril about middle of bill, placed near culmen. Head moderately large; neck very large, permitting the skin to pass over the head of the dead bird. Tail of eighteen feathers, stiff, narrow, and pointed, with large shafts. Tarsus half as long as middle toe and claw. Feet very large, outer toe longer than middle. Wings short.

Only one species of this genus, out of the twelve or fourteen recognized by ornithologists, is found in North America, the well-known Ruddy Duck, with many aliases in different localities. It is generally distributed throughout our Continent, going at times as far south as northern South America. It is a sprightly little bird with some rather comical habits.

GENUS NOMONYX

(Greek νόμος, *nomos*, law + ὄνυξ, *onux*, nail).

Nomonyx, Ridgw. Proc. U. S. Nat. Mus., vol. iii., 1880, p. 15. Type *Anas-dominica*, Linn.

Characters similar to Erismatura, but the nail nearly all seen from above, and, although hooked, does not bend backward. Tail composed of narrow graduated pointed feathers with stiffened shafts, and more than half as long as wing. Bill narrower for its length than is that of the allied genus. Outer toe shorter than the middle toe.

There is only one species included in this genus, a native of tropical America straggling into eastern North America, within whose boundaries a few examples have been captured. In full plumage the male is a very handsome bird, and rather smaller in size than the common Ruddy Duck.

SUBFAMILY MERGINÆ.

THE MERGANSERS.

This subfamily possesses three genera, and about nine species, and is represented in nearly every part of the world. They are generally known as the Saw-bill, or

Fish Ducks, and are not regarded as very desirable for the table, the flesh being generally impregnated with the flavor of fish, which is their principal food. They are birds of handsome plumage, with hues from delicate salmon tints to rich metallic greens. Some of the species breed in trees, and all are fond of frequenting secluded places, and keep much about the borders of marshes and tidal creeks when upon the coast, and are rarely seen on broad stretches of water. They fly very rapidly and are expert divers, and destroy immense numbers of small fish. Their peculiarly formed bill is apt to attract the attention of the most indifferent observer.

KEY TO THE GENERA.

A. Bill long, narrow, hooked.
  a. Culmen longer than tarsus.

    a'. Serrations of maxilla inclined backward.   GOOSANDER. RED-BREASTED MERGANSER. *Merganser.*

    b'. Serrations of maxilla not inclined backward.   HOODED MERGANSER. *Lophodytes.*

  b. Culmen shorter than tarsus.   SMEW. *Mergus.*

Of the first genus there are about seven species recognized, but two only are natives of North America, viz., the Goosander and the Red-Breasted Merganser. The second contains but one species,—the beautiful Hooded Merganser,—restricted to North America, very occasionally straying to Europe; while the third has the attractive Smew, an Old-World species included in our fauna on very slight grounds; the female, it is claimed, having been twice taken within our boundaries, the male never.

GENUS MERGANSER

(Latin *mergus*, a diver + *anser*, a goose).

Merganser, Briss. Orn., vol. vi., 1760, p. 230. Type *Mergus merganser*, Linn.

Culmen longer than tarsus; serrations of maxilla and mandible tooth-like, inclining backward. Bill long, narrow, tip hooked.

The species of this genus are large birds, the males with iridescent hues on the heads and necks. During the breeding season, when possible, they seek lakes and rivers within the forests and rear the young amid their solitudes. Flesh fishy and unpalatable.

KEY TO THE SPECIES.

A. Head and neck greenish black, metallic.
  a. No white collar on neck; under parts uniform. — GOOSANDER. *M. americanus.* ♂
  b. White collar on lower neck; under parts not uniform. — RED-BREASTED MERGANSER. *M. serrator.* ♂

B. Head and neck tawny brown.
  a. Distance between nostril and nearest feather at base of bill GREATER than height of maxilla at base. — GOOSANDER. *M. americanus.* ♀
  b. Distance between nostril and nearest feather at base of bill LESS than height of maxilla at base. — RED-BREASTED MERGANSER. *M. serrator.* ♀

GENUS LOPHODYTES

(Greek λόφος, *lophos*, a crest + δύτης, *dutes*, a diver).

Lophodytes, Reichenb. Syst. Av., 1852, pl. ix. Type *Mergus cucullatus*, Linn.

Culmen longer than tarsus. Serrations of bill blunt, not inclined backward.

Only one species of this genus is known, confined to North

America, the beautiful Hooded Merganser, noted for the expansive crest of black and white exhibited by the male, and from which it takes its name.

GENUS MERGUS

(Latin *mergus*, a diver).

Mergus, Linn. Syst. Nat., vol. i., 1766, p. 207. Type *Mergus albellus*, Linn.

Culmen shorter than tarsus.

A single species is included in this genus, a native of the Old World, of doubtful occurrence in North America, and popularly known as the Smew. It has a very attractive plumage of black and white.

L'ENVOI.

The history is finished, the self-sought task is done,
The tale is told of creatures wild and free;
Of a tribe that's swiftly passing, its course now nearly run,
Leaving for posterity naught save a memory.
We have heard the bell-like cry
Sounding faintly in the sky,
Of feathered squadrons speeding on their way;
We have watched the sportive broods
In the Arctic solitudes,
Where night was followed by an endless day.
We have known them in their glory, in the pride of numbers strong,
Now we see them gathering in a feeble company,
We have heard the waters echo to the music of their song,
Now we listen to the silence born of river, lake, and sea.

Nevermore in serried ranks, from fierce Atlantic's shore,
Across our wide domain to Pacific's tranquil sea,
The fowl will cloud the heavens, but the cry of "Nevermore,"
Shall echo to the limits of Ages yet to be.

INDEX.

ACADEMY OF NATURAL SCIENCES OF PHILADELPHIA, 27, 42
Æthyia, 284, 286
" americana, 59
" ferina, 286
Æx, 273
" galericulata, 87, 273
" sponsa, 90
Africa, North, 50, 98, 144, 178
" South, 282
Agattu Island, 76
Aix, 273
Alaska, Coast of, 19, 58, 84, 128, 188, 206, 220, 232, 271
Alaska, Territory of, 28, 35, 38, 45, 46, 52, 57, 72, 73, 74, 77, 84, 86, 122, 126, 128, 130, 136, 140, 142, 154, 160, 162, 164, 165, 169, 170, 180, 182, 184, 190, 196, 197, 198, 203, 212, 217, 219, 229, 230, 234, 235, 237, 240, 245, 249, 254, 257, 292, 293
Alaskan Peninsula, 199, 216, 217, 229, 232
Albemarle Sound, 40
Aleutian Islands, 19, 35, 46, 52, 54, 55, 69, 71, 72, 74, 75, 76, 77, 116, 134, 136, 160, 176, 184, 188, 189, 196, 201, 203, 206, 212, 217, 219, 220, 229, 232, 245, 249, 252, 271

America, 27
" Arctic, 27, 44, 171
" Central, 93, 136, 142
" Northern, 108, 203
" Northern South, 93, 130, 243
America, South, 93, 128, 132, 136, 242, 243
America, Western, 38, 70, 86, 133
Anas, 275, 277
" acuta, 279
" anser, 270
" bernicla, 271
" boschas, 104, 277, 278
" breweri, 103
" canagica, 270
" casarca, 276
" clangula, 288
" clypeata, 288
" crecca, 281
" cygnus, 265
" dominica, 296
" ferina, 286
" fuligula, 186
" fulvigula, 110, 278
" fulvigula maculosa, 112, 277, 278
Anas, glacialis, 290
" histrionicus, 290
" jamaicensis, 295
" mollissima, 293

Anas, nigra, 291
" obscura, 108, 278
" olor, 265
" penelope, 278
" querquedula, 280
" rufina, 285
" stelleri, 291
" strepera, 278
" valisneria, 285
Anatidæ, Family, 263, 264, 288
Anatinæ, Subfamily, 263, 274, 282
Anderson River, 45, 46, 57, 84, 85, 86, 165, 190, 230
Anderson River, Lower, 212
Anser, 268, 270
" albifrons, 48
" fabalis, 50
" hyperboreus, 269
" rossii, 269
Anseres, Order, 263
Anserinæ, Subfamily, 263, 267
Anthony, A. W., 198
Arctic Circle, 176
" Coast, 28, 190, 217
" Ocean, 19, 39, 46, 71, 74, 86, 118, 120, 170, 186, 192, 201, 225, 229, 230
Arctic Regions, 19, 27, 30, 31, 41, 42, 45, 52, 57, 68, 84, 100, 104, 113, 114, 122, 138, 147, 159, 162, 164, 165, 178, 182, 189, 192, 199, 204, 214, 227, 234, 235, 293
Arctic Sea, 35, 45, 57, 68, 70, 85, 118, 135, 160, 169, 175, 178, 184, 188, 201, 206, 212, 217, 234
Arctonetta, 284, 292
" fischeri, 220

Argentine Republic, 96, 132, 243
Aristonetta, 284, 285
Aristonetta valisneria, 152
Asia, 31, 50, 97, 259, 260, 265
" Northeastern, 292
" Southern, 98
Atkha Island, 69, 134, 249
Atlantic Coast, 21, 30, 33, 34, 35, 38, 40, 43, 45, 48, 70, 76, 78, 80, 83, 84, 85, 122, 124, 134, 142, 172, 175, 176, 196, 201, 206, 207, 213, 214, 222, 223, 225, 227, 234, 235
Atlantic Ocean, 19, 28, 39, 41, 45, 57, 68, 100, 117, 118, 135, 170, 203, 212
Auk, Great, 220
Australia, 265, 282, 286
Automniere, 128
Aythya, 286

Baedeker, 198
Baird, S. F., 271
Bald Crown, 120
" Eagle, 151
" Pate, 116, 118, 120
Barren Grounds, 28, 39, 124, 190, 212
Bec Scie, 254
Behring Islands, 217
" Sea, 19, 46, 55, 188, 206, 216, 220, 229, 232
Behring Straits, 45, 52, 55, 201, 234
Belden, Mr., 196
Bellot's Straits, 80
Berlin, 97
Bernicla, 271
Bermuda Islands, 204, 245
Bishop, Dr. L. B., 286
Black Head, 164

Black Head, Big, 160, 162, 286, 287
Black Head Creek, 164
" Little, 160, 161, 164, 165, 170, 287
Black Head, Ring-billed, 169
" Ringed-neck, 169
Black Neck, 162
Blanc Sablon, 172
Blue Bill, Big, 161
" Little, 164, 170
Blue Peter, 61, 192
Boardman, G. A., 170
Booby, 231
Branchier, 87
Brant, 78
" Blue, 34
" Gray, 47
" Pied, 47
" Prairie, 47
" Speckled, 47
" White, 37, 41
Branta, 271
" bernicla, 83, 272
" Black, 80, 83, 84, 85, 86, 272
Branta canadensis, 68, 77, 272
Branta canadensis, hutchinsi, 70, 77, 272, 273
Branta canadensis, minima, 77, 272
Branta canadensis, occidentalis, 73, 79, 272
Branta leucopsis, 79, 272
" nigricans, 86, 272
Brazil, Southern, 96
Brent, 82, 271
Brenthus, 271
Brewer, T. M., 196
Bristle Tails, 239
British Columbia, 133

British Islands, 31, 78, 134, 160, 217, 251, 257
British Museum, 259
" Provinces, 178
Broad Bill, 164, 169
" Bastard, 169
" Bay, 161, 162
" Big, 162
" Little, 161, 164, 165, 166, 170, 177
Broad Bill, River, 164
" Saltwater, 162
" Small, 162
Broady, 142
Brooklyn, 173
Brownsville, 242
Buffle Head, 63, 177, 184, 185, 238
Bull Head, 176
" Neck, 151, 239
Butter Ball, 184, 185
" Box, 184

CACCAWEE, 191
Cairina moschata, 104 (note)
Calais, 155, 170
Calaveras, 196, 198
California, Gulf of, 128
" Lower, 86, 201, 204, 214
California, Northern, 52
" State of, 19, 29, 35, 43, 55, 69, 72, 73, 74, 76, 77, 84, 95, 96, 116, 117, 130, 132, 147, 159, 188, 192, 195, 196, 199
California, Southern, 35, 38, 44, 93, 206, 212, 249
Camptolæmus, 284, 287
" labradorius, 175
Canada, 106

Canard cheval, 151
" français, 100
" gris, 113
" noir, 106, 169
" " d'Été, 111
" violin, 158
Canvas, 151
" Back, 118, 147, 148, 149, 150, 151, 154, 158, 161, 165, 186, 284, 285, 286
Canvas Back, Royal, 147
Canvas Backs, 149, 152, 157, 158
Cape Hatteras, 40
" St. Lucas, 48
Carolina, North, State of, 19, 20, 21, 40, 78, 87, 96, 104, 116, 117, 124
Carolina, South, State of, 177
Casarca, 275, 276
" casarca, 98
Charitonetta, 284, 289
" albeola, 186
Chaulelasmus, 275
" streperus, 114
Chen, 268
" cærulescens, 34, 269
" hyperboreus, 38, 269
Chesapeake Bay, 20, 21, 149, 152, 202, 214
Chili, 132, 133, 243
China, 47, 98, 134, 160, 162, 192, 206, 251, 160, 274
Churchill River, 33, 43, 58
Clangula, 283, 289
" clangula, 178, 289
" islandica, 102, 289
Cockawee, 191
Cock Robin, 254
Colorado, State of, 57, 113, 123, 133, 136, 180, 181, 182, 198

California, Northern, 245
Columbia River, 132, 142, 240
" " Upper, 198
Commander Island, 27, 45, 55, 140, 184, 206, 217, 233
Copenhagen, 50, 97
Coot, 214, 239
" Bay, 203, 208
" Black, 208
" Booby, 239
" Broad Bill, 239
" Brown, 203, 208
" Bumble Bee, 239
" Butter-billed, 208
" Butterboat-billed, 203
" Gray, 203, 208
" Horse Head, 203
" Hollow-billed, 203, 208
" Skunk Head, 203
" Spectacle, 203
" Whistling, 208
" White-winged, 212
Coots, 202, 203, 212
Coppermine River, 225, 229, 232
Corpus Christi, 111
Cotton Head, 254
Coues, Dr. E., 294
Coween, 191
Cuba, 39, 41, 48, 90, 104, 118, 120, 126, 136, 138, 175, 178, 184, 186, 237, 254, 257, 289
Currituck Sound, 19, 24, 40, 78, 87, 239
Cygne, 27, 29
Cygninæ, Subfamily, 263, 264, 265
Cygnus, 32, 265, 266
" columbianus, 27, 267
" cygnus, 32, 265, 266
" buccinator, 30, 267
" olor, 31

DABCHICK, 237
Dafila, 274, 275, 279
" acuta, 126
Dakotas, The, 30
Dall, W. H., 28, 54, 69, 164, 180, 216, 229
Delaware, State of, 172, 223
Dendrocygna, 274, 275, 276
" autumnalis, 93, 276
Dendrocygna fulva, 96, 276
Dipper, 184, 186
" Broad Bill, 239
Diver, Hell, 237
" Ruddy, 239
" Saw Bill, 254
Dos Gris, 158, 162
Dresser, H. E., 98, 195
Duck, Acorn, 87
" Black, 106, 107, 109
" " English, 106
" Braminy, 98
" Brewer's, 103
" Buffle Head, 184, 284, 289
Duck, Creek, 113, 278
" Dusky, 102, 106, 176, 278
" Dusky, Florida, 109, 111, 278
Duck, Eider, 294
" Fiddler, 294
" Fiddler, Yellow-bellied, 95
Duck, Fish, 245, 249, 297
" Fulvous Tree, 95, 276
" German, 113
" Gray, 100, 113
" Harlequin, 195, 196, 197, 198, 284, 290, 291
Duck, Heavy-tailed, 239
" Horse, 151

Duck, King, 293
" Labrador, 172, 220, 284
" Long-legged, 92, 95
" Long-tailed, 188, 191, 284
Duck, Mandarin, 272, 274
" Masked, 242, 295
" Mottled, 111, 278
" Muscovy, 104
" Noisy, 191
" Painted, 197
" Pied, 172
" Raft, 156, 159, 164
" Raft Red-headed, 159
" Ringed Neck, 169, 287
" Rock, 197
" Ruddy, 63, 237, 239, 242, 243, 285, 295, 296
Duck, Rufous-crested, 144, 283, 285
Duck, Rufous long-legged, 95
" Sand Shoal, 172
" Scaup, 160
" " Big, 165
" " Lesser, 164
" Skunk, 172
" Spectacle, 203
" Spirit, 176, 179, 184
" Steller's, 216, 217, 284, 291
Duck, Stock, 100
" Summer, 87
" Summer Black, 111
" Surf, 202, 203
" Surf Black, 214
" Surf White-winged, 214
" Swallow-tailed, 191
" Tree, Black-bellied, 92, 95, 276
Duck, Tufted, 169
" Velvet, 214

Duck, Wheat, 120
" White-faced, 128
" Wild, 100, 108
" Wild, Common, 277
" Wood, 87, 88, 89, 90, 263, 264, 273, 274
Ducks, 267
" Eider, 284
" Fresh Water, 263, 274, 277, 282
Ducks, Golden-Eye, 284
" Salt Water, 274
" Saw-billed, 245, 264
" Sea, 263, 274, 282
" Scaup, 158
" Spine-Tail, 263, 295
" Surf, 284, 291
" Tree, 275, 276
" Wood, 254

EIDER, 225, 226
" American, 222, 223, 224, 229, 232, 294
Eider, Common, 222, 224, 225, 227, 229, 233, 236, 294
Eider, Fischer's, 219, 235, 284, 293
Eider, King, 216, 232, 234, 235, 293, 294
Eider, Pacific, 216, 229, 231, 232, 294
Eider, Spectacled, 219, 235, 293
Eiders, 232, 234, 291, 293
Egypt, 31, 48
Egyptians, 48
England, 100
Eniconetta, 291
Erionetta, 294
Erismatura, 295
" jamaicensis, 240

Erismaturinæ, Subfamily, 263, 295
Eskimo, 32, 33, 54, 172, 231, 232, 235
Europe, 31, 50, 90, 144, 152, 182, 203, 204, 215, 251, 257, 259, 265, 289, 297
Europe, Central, 144
" Eastern, 98
" Northern, 50, 190, 217, 225, 260, 285
Europe, Southern, 50, 97, 98
Exanthemops, 268, 269
" rossii, 44

FALKLAND ISLANDS, 132, 133
Fanning Islands, 278
Faröe Islands, 78
Fiebig, Mr. Charles, 52
Fielden, Captain, 80
Finnish Lapland, 31
Fisherman, 249
Flocking Fowl, 164
Florida, State of, 106, 108, 109, 132, 133, 188, 192, 201, 203, 249, 254
Florida, Peninsula of, 109
Formosa, Island of, 274
Forrester, Frank, 214
Fort Anderson, 124
" Albany, 33
" Prince of Wales, 58
" Tejon, 93
Franklin Bay, 27, 85
Fulica americana, 61, 192
Fuligula, 284, 286
" affinis, 167, 287
" collaris, 171, 287
Fuligulinæ, Subfamily, 263, 274, 282

Fulton Market, 173
Fundy, Bay of, 222

GADWALL, 113, 114, 124, 275, 277, 278
Galveston, 95
" Bay, 19
Garrot, Rocky Mountain, 180
Geese, The, 263, 267
" Brant, 78
" Canada, 273
" Laughing, 268
" Snow, Greater, 268
" White-fronted, 270
" White-fronted European, 270
Georgia, State of, 235
Giraud, J. P., 172
Godthaab, 32
Golden Eye, 176, 177, 178, 183, 289
Golden Eye, American, 288
" Barrow's, 180, 181, 289
Golden Eye, Common, 180, 182, 288
Golden Eye, European, 178, 288
Goosander, 245, 246, 247, 249, 250, 251, 254, 297, 298
Goose, Bailey, 37
" Bald-headed, 34
" Bar, 79
" Barnacle, 78, 272
" Bay, 68, 70
" Bean, 50
" Black-headed, 68
" Blue, 33, 34, 269
" Brant, 70, 80, 82, 272
" Cackling, 48, 52, 69, 74, 76, 272

Goose, Canada, 36, 57, 61, 62, 70, 71, 72, 271, 272
Goose, Canada, Lesser, 70
" Cravat, 68, 268
" Emperor, 52, 54, 55, 72, 268
Goose, Eskimo, 70, 86
" Flight, 70
" Gray, 47, 68
" " Small, 70
" Hutchins', 69, 70, 71, 85, 272
Goose, Laughing, 47
" Lidenna, 55, 72
" Lower Ground, 46
" Mud, 70
" Prairie, 70
" Red, 41
" Reef, 68
" Ring, 83
" Snow, 33, 34, 35, 39, 40, 41, 43, 46
Goose, Snow, Blue, 34
" " Greater, 39, 41, 268, 269, 270
Goose, Snow, Lesser, 35, 37, 43, 44, 268, 269
Goose, Snow, Ross', 43, 77, 268
" Texas, 41
" Tundrina, 46, 70
Goose, White-cheeked, 72, 77, 272
Goose, White-collared, 52
" " -fronted, 45, 46, 52, 270
Goose, White-headed, 34, 35
" Wild, 57, 62, 66, 67, 71
" Wild, Common, 271
" Little, 70
" Winter, 70
" Yellow-legged, 47

Grand Menan, Island of, 33
Gray Back, 158, 162
Greaser, 239
Great Britain, 48, 98, 178, 192, 259, 285, 286
Great Head, 176
Great Lakes, 201, 206, 212, 222, 223, 234, 235, 282
Great Slave Lake, 225, 237
Green Head, 100
Greenland, 31, 32, 45, 48, 50, 78, 80, 98, 134, 180, 182, 224, 225, 227, 251, 254, 257, 266, 277, 294
Greenland, Glacier Valley, 42
" West, 97
Greenland, North, 50, 97
" South, 32
Guatemala, 118, 120, 162, 164, 167, 169, 171, 237, 240
Guinea Hen, 55

HAIRY CROWN, 254
" Big, 249
Hairy Head, 254, 256
Harelda, 290
Havelda, 284, 290
" glacialis, 192
Hearne, 38, 43, 58, 137
Hemisphere, Eastern, 34, 79, 160, 195, 237, 266, 290, 292
Hemisphere, Northern, 140, 235, 293
Hemisphere, Western, 290
Heniconetta, 284, 291
" stelleri, 217
Hewitson, 198
Histrionicus, 284, 290
" histrionicus, 199
Honduras, 135, 138
Hooper, The Wild, 31

Horicon, Lake, 155
Hudson Bay, 28, 33, 34, 39, 43, 78, 80, 87, 90, 106, 136, 137, 206, 212, 237, 240
Hudson Bay Company, 259
Humboldt Bay, 52, 55

ICELAND, 31, 32, 78, 97, 98, 180, 182, 225, 251
Icy Cape, 234
Illinois, State of, 28, 116, 272
India, 98, 144, 260
Indies, West, 93, 128, 130, 164, 167, 169, 171, 242, 243
Iowa, 28, 30

JAMES BAY, 33
Japan, 31, 47, 48, 98, 122, 134, 178, 192, 195, 199, 206, 251, 259, 260, 274, 286
Jerdon, Dr. T. C., 98

KADIAK ISLAND, 216, 217
Kamchatka, 27, 55, 140, 217
Kansas, State of, 112, 130, 277
Kennicott, R., 86, 164
Kerguelen Island, 280
Kittitas, 198
Koshkonong Lake, 242
Kotzebue Sound, 160, 201, 206
Kurile Islands, 197, 198, 217
Kuskokwim River, 219, 220

LABRADOR, 33, 108, 172, 201, 213, 222, 223, 225
Lady, 197
La Fresnaye, Baron de, 271 (note)
Lake Champlain, 242
" Erie, 181
" Michigan, 188, 212

Lapland, 160
Larus philadelphia, 251
Laxa River, 195
Leggett, F. W., 24
Leucoblephara, 271 (note)
Leucoblepharon, 271 (note)
Leucopareia, 271 (note)
Liverpool Bay, 40, 85
Long Island, 40, 45, 78, 134, 172
Lophodytes, 297, 298
" cucullatus, 257
Lord, 197
Louisiana, 27, 29, 34, 37, 47, 87, 92, 95, 96, 106, 111, 113, 114, 125, 128, 142, 158, 162, 169, 176, 245, 254, 259

MACFARLANE, MR., 27, 45, 46, 85, 165, 230
Mackenzie River, 39, 41, 212
" " District, 136
Mayoum, 48
Maine, State of, 33, 34, 106, 155, 170, 180, 186, 196, 222, 225
Malden, 242
Mallard, 61, 100, 101, 102, 103, 104, 106, 107, 108, 141, 144, 151, 176, 275, 278
Mallard, Black, 106
" Dusky, 106
" Gray, 100
Manitoba, Province of, 123
Mareca, 275, 278
" americana, 120, 279
" penelope, 117, 279
Marionette, 184
Massachusetts, State of, 101, 172, 176, 178, 227, 242, 243
Matamoras, 242

Mazatlan, 95
Mediterranean, 259, 285
Merganser, 181, 249, 251, 254, 297, 298
Merganser, American, 245
" Americanus, 247, 298
Merganser, Buff-breasted, 245
" Hooded, 68, 89, 238, 254, 255, 297, 299
Merganser, Red-breasted, 249, 250, 251, 256, 297, 298
Merganser, Red-headed, 249
" serrator, 252, 263, 298
Merginæ, Subfamily, 263, 264, 296
Mergus, 297, 299
" albellus, 260, 299
" cucullatus, 298
" merganser, 298
Mesquin, 142
Mexico, 45, 48, 93, 96, 142, 161, 170, 178, 184, 186, 254, 257
Mexico, Gulf of, 19, 28, 29, 30, 33, 34, 35, 45, 69, 70, 87, 90, 92, 93, 100
Mexico, Western, 242
Minnesota, State of, 164, 169, 212, 237
Mississippi, River, 138, 190
" Valley of the, 19, 20, 29, 33, 34, 35, 38, 39, 45, 69, 70, 74, 77, 80, 86, 108, 128, 132, 133, 170, 237, 269
Missouri, State of, 96, 201, 206, 212, 214
Mit-huk, 232
Montana, State of, 44, 186
Mud Hen, 61, 192
Museum of Natural History,

The New York, 69, 116, 174, 181, 272
Museum, United States National, 197, 198, 265

NEARER ISLANDS, 206
Nelson, E. W., 49, 52, 74, 123, 164, 191, 196, 201, 207, 219, 220, 229
Netta, 283, 285
" rufina, 145
Nettion, 275, 280, 281
" carolinensis, 138, 281
" crecca, 135, 281
New Brunswick, Province of, 101
Newfoundland, 195, 199
New Jersey, State of, 140, 173, 175, 206, 234
New Orleans, 151, 259
New World, 45, 114, 195, 199, 234, 249, 270
New York City, 134, 144, 173
New York, State of, 180, 182, 196, 242
Nevada, State of, 95, 96
Nile, The River, 48
Nomonyx, 295, 296
Norfolk, 202
North America, 19, 28, 30, 45, 54, 57, 68, 78, 83, 87, 97, 100, 104, 106, 113, 118, 120, 124, 125, 130, 132, 134, 135, 136, 138, 140, 142, 144, 147, 152, 154, 159, 161, 162, 164, 167, 169, 170, 171, 174, 175, 178, 182, 186, 188, 192, 195, 198, 199, 201, 206, 212, 214, 217, 223, 225, 227, 234, 237, 240, 245, 247, 249, 254, 257, 259, 260, 264, 265, 266, 267, 269, 271, 273, 274, 276, 277, 280, 281, 285, 286, 288, 289, 290, 291, 292, 295, 296, 297, 298, 299
North America, Eastern, 79, 80
Norton Bay, 219
" Sound, 55, 85, 201, 217
Norway, 225
Nova Scotia, 78, 79
Nueces Bay, 111
" River, 95

ŒDEMIA, 283, 284, 291
" americana, 292
" carbo, 292
" deglandi, 214, 292
" perspicillata, 203, 292
Ogdensburg, 181
Ohio River, 188, 204
Oidemia, 291
Oie blanche, 37
Oie bleu, 34
Oie caille, 47
Old Granny, 161
" Injun, 191
" Molly, 191
" South Southerly, 188
" Squaw, 188, 189, 191, 192
" Squaws, 190, 290
" Wife, 191
" World, 31, 45, 47, 50, 78, 97, 114, 116, 122, 134, 135, 144, 147, 162, 178, 192, 195, 199, 224, 234, 249, 254, 259, 270, 274, 276, 279, 285, 286, 288, 294, 297, 299
Olor, 265, 266
" bewickii, 265
" buccinator, 265
" columbianus, 265
Oregon, State of, 35, 147

Outarde, 68
Ouzel, 195

PACIFIC COAST, 28, 30, 33, 34, 35, 38, 44, 55, 69, 72, 73, 80, 84, 128, 132, 147, 170, 201, 212, 214, 234, 235, 237, 246, 249, 271
Pacific Ocean, 19, 45, 67, 68, 70, 100, 117, 118, 135, 170, 176, 184, 203, 278
Panama, 104, 126
Patagonia, 133
Pearson, H. J., 195, 198
Peary, Lieut., 42
Pewaukee, Lake, 169
Pheasant, Water, 125, 254
Philacte, 268, 270
" canagica, 55
Pian Queue, 125
Pigeontail, 125
Pigeon, Wild, 88
Pintail, 118, 122, 123, 124
Plectropterinæ, 263, 273
Plongeur, 176
Poacher, 120
Pochard, American, 159
Point Barrow, 74, 84, 217, 219, 220
Polar Ocean, 84
Potomac River, 188
Printempsnierre, 128
Prybilof Islands, 54
Puckaway Lake, 69, 149, 262
Puffins, 197
Puget Sound, 196

QUERQUEDULA, 275, 280
" cyanoptera, 133, 280, 281
Querquedula, discors, 130, 133, 280, 281

RED HEAD, 118, 147, 150, 154, 155, 156, 157, 161, 284, 285, 286
Red Heads, 149, 156, 157, 158, 240
Reichenbach, 271
Richardson, 40, 57
Ridgway, R., 183
Ring Bill, 169
Ring-gaas, 83
Ring Neck, 170
Rio Grande, 92, 95
Rio Grande, Lower, 243
Rockies, 197
Rock River, 242
Rocky Mountains, 154, 180, 195, 199
Rook, 239
Ross, R. B,, 78
Ruddy, 239
Rupert House, 78
Russia, 216

SACRAMENTO, 95
" Valley of the, 155
Salt Lake Valley, Great, 43
Sarcelle, 138
Sasarka, 55
Saskatchewan, Valley of the, 136
Sauk Island, 216
Saw Bill, 249
" Little, 254
" Sea, 245
Scandinavian Peninsula, 97, 98
Scaup, Big, 162, 164, 167
" Little, 124, 160, 162, 164, 169, 170
Scaup, Big, Ringed-neck, 169
Schönherr, Mr., 271
Scie de mer, 245

Scolder, 191
Scoter, American, 206, 207, 212, 292
Scoter, Lake Huron, 214
" Surf, 201, 203, 207, 292
" Velvet, 214, 292
" White-winged, 201, 212, 292
Scoters, 201, 283
Scotland, 27
Seal Islands, 196
Sé-le-sen, 100
Semiché Islands, 76
Sennett, G. B., 111, 112
Sheldrake, 249
" Buff-breasted, 245
" Pied, 240
" Pond, 254
" Ruddy, 97, 275
" Swamp, 245, 252
Shelduck, 249
Shepard, C. W., 195, 196
Shoa, 98
Shoveler, 140, 275, 282
" Blue-winged, 142
" Mud, 142
" Red-breasted, 142
Shuffler, 164, 169
" Big, 162
" Ring-billed, 169
Shumagin Islands, 206, 216
Siberia, 78
Siberian Coast, 19, 52, 55, 217
Sierra Nevada, 195, 197, 199
Silverton, 198
Singley, J. A., 111
Sitka, 72, 73, 80, 201, 212
Smee, 125
Smew, 99, 259, 260, 297, 299
Snipe, 122
Somateria, 284, 293

Somateria, dresseri, 223, 294
" mollissima, 224, 227, 233, 294
Somateria, spectabilis, 235, 294
" v-nigrum, 232, 294
South Southerly, 191, 290
South South Southerly, 188
Spatula, 275, 282
" clypeata, 142
Speckle Belly, 47, 113
Speculum, xix
Spike Bill, 254
Spike Tail, 125
Spine Tail, 239
Spitzbergen, 78, 83
Spoonbill, 140, 142
Spreet Tail, 125
Sprig, 125,
Sprigtail, 119, 122, 125, 158, 189, 275
Sprigtail, American, 280
Squealer, 87
St. Croix River, 170
St. Lawrence, Gulf of, 180, 182, 196, 213, 227
St. Lawrence Island, 206, 216, 234
St. Lawrence, River, 181
St. Louis, 206, 212, 214
St. Michael's Island, 19, 46, 52, 55, 69, 72, 84, 201, 207, 212, 219, 234, 249
Stanislaus River, 196
States, Eastern, 128, 133
" Middle, 195, 199, 212
" New England, 213
" Northern, 114, 249, 252
" Northeastern, 101
" Northwestern, 152
" South Atlantic, 101
" Southern, 104, (note)

States, Western, 170
Steel Head, 239
Stejneger, L. J., 184, 233, 265, 266
Stewart Island, 55, 101
Stick Tail, 239
Stiff Tail, 239
Subarctic Coast, 217
Sundevall, 271
Swaddle Bill, 142
Swan, American, 26, 28
" Australian, 264
" Bewick's, 32
" Mute, 31, 32
" Trumpeter, 20, 26, 28, 29, 267
Swan, Whistling, 20, 28, 29, 266, 267
Swan, Whooping, 32, 266, 267
Swans, The White, 265, 266
Swinhoe, R., 206

TAPKAN, 234
Teal, 77
" American Green-winged, 135, 136, 281
Teal, Blue-winged, 280, 281
" European, 134
" " Green-winged, 135, 281
Teal, Green-winged, 128, 130, 134, 136, 137, 275
Teal, Mud, 138
" Red-headed, 138
" Salt-water, 239
" Scotch, 184
" Spoonbill, 142
" Summer, 128, 132, 280, 281
" Winter, 138
Teals, Green-winged, 275
" Blue-winged, 275

Texas, State of, 45, 86, 93, 95, 96, 106, 111, 118, 140, 142, 188, 192, 237, 242
Texas, Eastern, 112
" Southern, 133
" Western, 29
Trinidad, 212
Trumpeter, The, 266
Turkestan, 285
Turner, L. M., 54, 55, 74, 76, 134, 164, 197, 203, 219, 229, 249

UNALASKA, ISLAND, 55, 74, 77, 188, 196, 216, 245
United States, 20, 28, 35, 43, 57, 60, 69, 70, 77, 80, 92, 106, 116, 117, 120, 124, 128, 129, 132, 136, 138, 147, 148, 160, 164, 167, 169, 171, 191, 203, 234, 242, 247, 251, 252, 254, 255, 259, 277
United States, Eastern, 145
" Northern, 30, 45, 52, 68
Uppernavik, District of, 97
Utah, 106, 180, 182

VANHOFFEN, DR., 97
Vancouver Island, 147
Vineyard Island, 172

WALKER, DR., 80
Washington Market, 173
" State of, 35, 198
Waukareen, 234
Wavey, Blue, 34, 269
" Common, 41, 43
" Horned, 43
Waveys, 34
Weaser, 245
Webster, Daniel, 172

Welch Drake, 113
Wenatchee, 198
Wenge, Mr., 97
Whippler, 176
Whistler, 176, 177, 178, 180
Whistle Wing, 176
White Back, 151
White Sea, 78
Widgeon, 89, 113, 114, 118, 119, 120, 124, 156, 275, 277, 280
Widgeon, American, 114, 116
" Bald-faced, 120
" European, 116, 279
" Gray, 113
" Green-headed, 120
" Wood, 87

Wilson, A., 172
Wisconsin, Lakes of, 251
Wisconsin, State of, 69, 74, 77, 136, 149, 155, 169, 188, 212, 222, 234, 242, 243
Wrangel Land, 84

XANTUS, MR., 93

YUKON, DELTA OF THE, 69, 72
" Fort, 28, 33, 45
" River, 19, 35, 52, 54, 55, 57, 74, 84, 86, 122, 140, 147, 160, 164, 177, 180, 196, 201, 229, 232

ZAN-ZAN, 120

THE END.

UNIFORM WITH THE WILD FOWL

NORTH AMERICAN SHORE BIRDS

A Popular History of the Snipes, Sandpipers, Plovers, etc., inhabiting the beaches and marshes of the Atlantic and Pacific coasts, the prairies, and shores of the inland lakes and rivers of the North American Continent. With 74 fine full-page plates specially drawn for this work by Mr. Edwin Sheppard.

THE GALLINACEOUS GAME BIRDS OF NORTH AMERICA

Including the Partridges, Grouse, Ptarmigan, and Wild Turkeys; with accounts of their dispersion, habits, nesting, etc., and full description of the plumage of both adult and young, together with their popular and scientific names. A book written both for those who love to seek these birds afield with dog and gun, as well as those who may only desire to learn the ways of such attractive creatures in their haunts, with 46 fine full-page plates specially drawn for this work.

These popular ornithological books by Prof. Daniel Giraud Elliot are acknowledged to be the standard works on this subject.

Crown Octavo, Ornamental Cloth, $2.50 Each
Large-paper Edition, Limited to 100 Copies Signed by the Author
$10.00 Net

The Nation: "Scientific phraseology has been, as far as possible, carefully avoided. . . Mr. Elliot's condensed, well-written, and thoroughly trustworthy biographies will prove most welcome," etc.

Forest and Stream: "Naturalists, sportsmen, and bird-lovers generally are to be congratulated on the appearance and character of this volume. . . . It is especially admirable for its simplicity and directness and for the very high quality of its many illustrations."

London Field "Mr. Elliot is to be congratulated upon his artist. The illustrations by Mr. Edwin Sheppard are soft and delicate in outline; they convey an accurate impression of the species represented."

Commercial Advertiser: "One of the most interesting and valuable books of its kind that have been written for some time."

FRANCIS P. HARPER

17 East 16th Street   NEW YORK

www.ingramcontent.com/pod-product-compliance
Lightning Source LLC
Chambersburg PA
CBHW020526300426
44111CB00008B/562